The Mathematical Olympiad Handbook

The Mathematical Olympiad Handbook

An introduction to problem solving
based on
the first 32 British Mathematical Olympiads
1965–1996

A. GARDINER

School of Mathematics
University of Birmingham, UK

OXFORD
UNIVERSITY PRESS

OXFORD

UNIVERSITY PRESS

Great Clarendon Street, Oxford OX2 6DP

Oxford University Press is a department of the University of Oxford.
It furthers the University's objective of excellence in research, scholarship,
and education by publishing worldwide in

Oxford New York

Athens Auckland Bangkok Bogotá Buenos Aires Calcutta
Cape Town Chennai Dar es Salaam Delhi Florence Hong Kong Istanbul
Karachi Kuala Lumpur Madrid Melbourne Mexico City Mumbai
Nairobi Paris São Paulo Shanghai Singapore Taipei Tokyo Toronto Warsaw

with associated companies in Berlin Ibadan

Oxford is a registered trade mark of Oxford University Press
in the UK and in certain other countries

Published in the United States
by Oxford University Press Inc., New York

© A. Gardiner, 1997

The moral rights of the author have been asserted

Database right Oxford University Press (maker)

First published 1997
Reprinted 1999, 2000

A catalogue record for this book is available from the British Library

Library of Congress Cataloging in Publication Data
Gardiner, A. (Anthony), 1947-
The Mathematical Olympiad Handbook: an introduction to problem
solving based on the first 32 British mathematical olympiads
1965–1996 / A. Gardiner.
1. Mathematics–Problems, exercises, etc.   2. British Mathematical
Olympiad.   I. Title.
QA43.G345   1997   510'.76—dc21   97-17647

ISBN 0 19 850105 6

Printed in Great Britain by Biddles Ltd., Guildford and King's Lynn

...a mathematical problem should be difficult in order to entice us, yet not completely inaccessible lest it mock our efforts. It should be a guidepost on the tortuous path to hidden truths, ultimately rewarding us by the satisfaction of success in its solution.

David Hilbert, 1900

Preface

This is unashamedly a book for beginners. Unlike most Olympiad problem books, my aim has been to convince as many people as possible that Mathematical Olympiad problems are for *them* and not just for some bunch of freaks. I have tried never to use one slick, unmotivated step where three down-to-earth steps would do just as well. Once you are convinced that these problems are do-able and are worth the effort required to solve them, you will be ready to move on to other books of Olympiad problems—some of which are listed in the section 'Some books for your bookshelf'.

Who is the book for?

In mathematics, as in music and sport, there are many youngsters who are capable of performing at a much higher level than is required by the ordinary school curriculum. While many youngsters have the necessary *potential*, the final ability is not God-given: it has to be developed—and that requires effort and commitment. Such students may remain in their peer groups, but need higher goals to aim at. This book is for such students aged 15 or 16 and above.

The book presumes an initial commitment on the part of you—the reader. At first, this commitment need go no further than a basic technical competence, combined with a willingness to struggle on when faced with a challenging mathematical problem. But sooner or later you will have to make the effort to follow up the mathematical ideas which lie behind these problems, and start reading other books to master some unfamiliar mathematics on your own. Reader-friendly mathematics books can be hard to find: to guide you, the section 'Some books for your bookshelf' contains a short list of books which you may find helpful, together with a very brief summary of the mathematics covered by each one.

What is in this book?

The main part of this book contains problems from the first 32 British Mathematical Olympiad (BMO) papers 1965–96, starting with 1996 and working backwards. This is followed by a section in which the problems are considered in turn, with each problem being followed by a page or so of text. The first part of this text should be read only after you have had a good go at the problem for yourself. You should then use these fresh ideas for a second attack on the problem. Only later—whether or not you have succeeded in solving the problem—should you work through the whole 'Outline solution'. If you failed to solve the problem, the text should convince you that there is at least one 'obvious' line of attack which you could have discovered for yourself. If on the other hand you managed to solve the problem on your own, the text may still open your eyes to some things you never thought of. However, the text itself does not present a complete solution. To obtain a complete solution you must fill in for yourself the details in the 'obvious' approach outlined in the text.

Most of the problems in this book can be solved in many different ways. The first correct solution that one finds to a problem is almost never the simplest. It is only *after* one has succeeded in solving a problem that one can sometimes look back and see that there is in fact a much easier solution—if only one looks at things in the right way!

The papers from the years 1965–74 are different in style, in that each contains ten or eleven (not necessarily original) problems—including the occasional mechanics problem. I have included the statements of the problems both for completeness and because they include many nice examples; however, I have not provided any discussion, hints, or outline solutions for these problems.

Most Olympiad problems do not require advanced mathematics for their solution. However, they do require familiarity with some basic ideas which receive little attention in most secondary school curricula. The section 'A little useful mathematics' provides a brief introduction to the simplest and most basic topics. You may find it helpful to work through this section for a first time before tackling the problems in the rest of the book. Later, as you meet problems that use some of these ideas, you may need to work through some parts of the section in more detail.

Why was the book written?

Textbook exercises and examination questions are not meant to be particularly demanding: textbook exercises are intended to help all students practise some standard technique, and exam questions are supposed to give ordinary students an opportunity to show what they have learned. Thus both these types of problems should be relatively straightforward for the large number of students who are capable of performing at a higher level. This does not mean they should be skipped! Indeed, it is perhaps more important for able students than for others that they receive a solid grounding in basic technique. But this is only a beginning: such students need a regular diet which goes beyond mere textbook exercises and examination questions.

This book should be especially useful to those who are preparing to take part in Olympiad-type competitions. Taking part in competitions, and working through old competition questions, provides one very valuable source of extension and enrichment. Young musicians and sportsmen understand perfectly well that the harder one practises or trains, the more likely one is to benefit from performing or competing at a high level. In contrast, mathematics competitions are frequently tackled without any preparation at all! This can result in an experience that turns out to be memorable for all the wrong reasons; instead of catching an exciting glimpse of new mathematical horizons, the poorly prepared student can sometimes be rendered helpless by a kind of mathematical stage-fright.

Take-home competitions, which allow students one to three *weeks* to solve five or six problems, can sometimes get round this lack of preparation. Faced with take-home problems that are sufficiently accessible, talented but poorly prepared students can perform creditably and can learn a lot by working hard and training 'on the job'. (Examples of this approach working successfully on a large scale in the UK include the Scottish Mathematical Challenge, the Merseyside Challenge, and the Birmingham-based BUMMPS and Junior PEST.) However, it is unreasonable to expect talented but poorly prepared students to pull themselves up by their own bootstraps in a timed event which allows them just one to three *hours* to solve five or six challenging, non-standard problems.

The fault here lies not with the timed, written format! Timed, written papers have an invaluable role to play: they stress the importance of thinking both creatively and quickly, and force students to present simple, clear solutions. For administrative reasons, any large national competition is almost bound to involve a timed paper. Moreover, the kind of questions traditionally set in such competitions have tremendous potential for introducing talented students to an accessible, yet completely new mathematical world.

In the past it has been very difficult for ordinary schools to help their best students prepare for, and hence benefit from, timed written papers of this kind. As a result many very able students have performed in a way which may sometimes have been discouraging rather than stimulating. This book should help *all* schools and colleges to introduce able students in their last two or three years at school to a new world of challenging mathematical problems.

The book is unlikely to succeed unless it also persuades *teachers* that these Olympiad problems, though by no means easy, are more accessible than they may have previously imagined. The experience of tackling and solving (or of failing to solve and being annoyingly frustrated by) hard mathematical problems is an eternal source of delight and disappointment. I hope that this little book will open up this world of delights and disappointments to many who would otherwise have looked on bemusedly from the sidelines.

While I accept full responsibility for the structure of the book and the style of the text, the contents owe much to many people. When the project was first begun back in 1991, much encouragement and support was given by Richard Higson and Chris Nash. The problems themselves are due to a number of other people, whose individual contributions have been blurred by the passing of time (in the early years, stalwart work was done by Margaret and Walter Hayman; in the middle years, the mainstays were Robert Lyness and David Monk; more recent papers stem from the work of the Problems Group of the British Mathematical Olympiad Committee). The section 'A little useful mathematics' had its roots in the original tentative draft by John Deft of a booklet designed to help students prepare for the BMO. Dozens of others have contributed in many different ways: in particular, Christopher Bradley, Zad Khan, Adam McBride, David Monk, and Robert Pargetter commented in detail on various drafts, and helped to reduce the number of errors. I can only hope that each one of them will feel that this presentation remains faithful, at least in part, to what they were trying to achieve. To all of them, I express my thanks.

Finally, to those readers who are entering this mathematical world for the first time, may I say 'Bon voyage'.

Contents

Part III Hints and outline solutions

Problems and problem solving

However unapproachable these problems may seem to us and
however helpless we stand before them, we have, nevertheless,
the firm conviction that their solution must follow by a finite
number of purely logical processes. [...]
*   This conviction of the solvability of every mathematical*
problem is a powerful incentive to the worker. We hear within
us the perpetual call: There is the problem. Seek its solution.
You can find it by pure reason; for in Mathematics there is no
ignorabimus.

David Hilbert, 1990

David Hilbert (1862–1943) was one of the most outstanding mathematicians of the modern era. At the International Congress of Mathematicians in Paris in 1900, he presented 23 major research problems which he felt would be important for the development of mathematics in the twentieth century. These problems all seemed very hard, but in bringing them to the attention of other mathematicians Hilbert felt that the need to stress that this should not be used as an excuse to put off trying to solve them.

Hilbert's judgement that these problems would play a central role in the mathematics of the present century was remarkably accurate. But the most interesting thing for us here is his rallying call: However unapproachable these problems may seem at first sight, and however helpless we stand before them, we have the firm conviction that their solution must be possible by purely logical processes. There is the problem. Seek its solution. You can find it by pure reason.

When faced with an unfamiliar and apparently very difficult mathematical problem, there is always the temptation to imagine that it is *too* hard, that progress towards a solution requires some trick or technique that we have not yet learned, and that the solution is therefore beyond our powers. This defeatist view is all the more plausible because it must sometimes be true! It is nevertheless misguided. Let me try to explain why.

During the nineteenth century it became clear that the more scientists discovered about nature, the more they realized how *little they knew*, and that one could never hope to discover the *whole truth*. This realization was summed up by one philosopher, Emil du Bois Reymond, in the phrase 'ignoramus et ignorabimus'—ignorant we are and ignorant we shall remain. This catch phrase certainly caught on! And as the new century dawned, David Hilbert felt it important to state as clearly as he could that *mathematics is different*. In mathematics, said Hilbert, we can tackle problems 'with the firm conviction that their solution must follow by a finite number of purely logical processes'. As if to underline his assertion, one of his 23 outstanding problems was solved almost immediately.

Hilbert was talking about mathematical *research*. However, his principle applies just as well to solving Olympiad problems. When mathematicians begin to explore a problem, they take it for granted that it must be possible to make progress towards a solution *using only the tools currently available*. Of course, they are sometimes wrong; but that is not the point. Mathematicians know perfectly well that, strictly speaking, the assumption is irrational (in that it cannot be justified logically, and is in general clearly false). *But it is an invaluable working hypothesis*. Though strictly illogical, the assumption that every mathematical problem can be solved has justified itself so often in practice that it has become a powerful conviction—one which is psychologically invaluable each time we experience that feeling of helplessness when trying to get to grips with a hard mathematical problem.

Faced with an unfamiliar mathematical problem, the mathematician— whether young or old—is like someone with a hopelessly small bunch of 'keys' who is trying to open up some fiendishly difficult Chinese puzzle box. At first glance the surface seems totally smooth, without a single visible crack. If you were not convinced that it is indeed a Chinese puzzle box, and that it can in fact be opened, you would soon give up. Knowing (or rather believing) that it can be opened, you may be willing to keep searching until you eventually begin to discern the slightest hint of a crack here and there. You still have no idea how the pieces are meant to move, or which of your 'keys' may help you open up the first layer of the puzzle. But by trying the most appropriate-looking keys in the most promising cracks, you eventually stumble on one that fits exactly, and the pieces begin to move.

In a good puzzle, success is never the result of pure chance. Indeed, once one has discovered the way in, one often realizes that it should have been obvious where to begin. But things usually become 'obvious' only in retrospect. If the puzzle is unfamiliar and well-made, success may require persistence, faith, and much time. Exactly the same is true of any good mathematics problem. So never give up too easily, and always be prepared to look back after solving a problem to see what you should perhaps have done differently. That is how you learn.

How to use this book

We must know, we will know.
David Hilbert, 1930

You have no choice but to tackle the problems in this book using the bunch of 'keys', or mathematical techniques, that you already know—no matter how limited they may be. It is also important to learn new tricks, and to revise old ones, as you go along. Thus you may choose to begin by working carefully through the section 'A little useful mathematics'. But you must not let this interfere with your basic hypothesis that *every problem has to be solved using only the techniques that you already know*.

In this book, the text at the beginning of the outline solution for each problem is meant to convince you that the problem could indeed have been solved using only the techniques that you know. *Do not read this text until you have tried really hard to solve the problem for yourself.*

Sometimes the text may not succeed in convincing you that you could have solved the problem on your own—although even then it should show how, using only the techniques that you know, you could have got further than you did. On other occasions the text will only convince you that you need to learn, or brush up, some bit of mathematics which is needed for that particular type of problem. When this happens you should first look back at 'A little useful mathematics'. If that does not help, I suggest you consult one of the books listed in 'Some books for your bookshelf'.

As a guide I suggest that you allow yourself at least one hour to work on each problem before consulting the relevant text in 'Hints and outline solutions'. Whenever possible you should give yourself time by 'sleeping on the problem' before looking at the text the next day. If you make a habit of this, you will discover the curious fact that, whenever you have been thinking hard about a problem, all sorts of useful ideas come into your head just as you begin to relax.

Part I
Background

A little useful mathematics

*...what is clear and easy to understand
attracts us, what is complicated repels us.*
David Hilbert, 1900

Introduction

The problems which are set in the British Mathematical Olympiad (BMO) are not all that hard. However, they can seem rather hard if you are not familiar with certain important ideas which are often skated over, or ignored altogether, in school mathematics. This section introduces some of the most basic of these ideas. If you work on it before tackling the rest of the book, you will learn a lot of useful mathematics and will get far more out of tackling the problems than if you simply try to solve the problems without any preparation. In particular, *make sure that you tackle all the exercises marked with a single asterisk* (∗). (Harder exercises, which you may well find too difficult on a first reading, are marked with a double asterisk (∗∗).) Whenever blanks are left in the text, you should make sure you know how to fill them in before reading further.

1. Numbers

Our real number system $\mathbb{R} \supseteq \mathbb{Q} \supseteq \mathbb{Z} \supseteq \mathbb{N}$ has several *layers*, like an onion. $\mathbb{R}$ denotes the set of all *real* numbers; including, for example,

$$-6\tfrac{1}{4}, \quad -3, \quad 1-\sqrt{2}, \quad 0, \quad \sqrt{2}, \quad \frac{\pi}{2}, \quad \sqrt[3]{10}, \quad e, \quad 1+\sqrt{3}, \quad 666.$$

If we remove the *irrational* numbers (those numbers which cannot be expressed as the ratio of two integers), we are left with the set $\mathbb{Q}$ of all *rational* numbers, or fractions; that is, numbers such as

$$-4, \quad -\tfrac{16}{5}, \quad 0, \quad \tfrac{1}{2}, \quad 1, \quad 1.414, \quad 2\tfrac{1}{7}, \quad \tfrac{61}{29}, \quad 3\tfrac{1}{3}.$$

Each rational number can be expressed in lowest terms as a quotient m/n, where m and n are whole numbers having no common factors and $n \neq 0$; for example, $1.414 = 1414/1000 = 707/500$. If we take only those rational numbers m/n with denominator $n = 1$, we obtain the set $\mathbb{Z}$ of *integers*:

$$\mathbb{Z} = \{\ldots, -4, -3, -2, -1, 0, 1, 2, 3, 4, 5, \ldots\}.$$

The *positive* integers

$$\mathbb{N} = \{1, 2, 3, 4, 5, 6, 7, 8, \ldots\}$$

are often called the *natural* numbers.

Whenever we are working with numbers, the kind of operations which are permissible, and the kind of statements and deductions which are valid, will depend on which kind of numbers we are using.

(1.1) Every real number x has a negative $-x$, which is also a real number. If x is a rational number, so is $-x$. If x is an integer, so is $-x$. But if x is a natural number, $-x$ is *not* a natural number.

(1.2) A real number x has an inverse $1/x$ provided that x is *non-zero*. If x is a (non-zero) real number, then $1/x$ is also a real number.

(∗) If x is a non-zero rational number, is $1/x$ always a rational number?

(∗) Suppose that x is a non-zero integer. When exactly is $1/x$ also an integer?

If a problem is restricted to *integers*, one has to be very careful about writing *fractions* (since the quotient of one integer by another will usually not be an integer), or about taking *square roots* (since, for example, when x is an integer, $\sqrt{x}$ will usually not be an integer).

One also has to be careful about 'cancelling' the common factor 'a' on each side of an equation of the form '$ax = ay$' to conclude '$\therefore x = y$'. Cancelling is a short way of 'multiplying both sides of the equation by $1/a$'—and this is possible only if a is *non-zero*. Thus, whenever one is tempted to cancel two expressions that *look* equal, one must always think twice. For example, given the equation '$6(x - 1)/x = x^2 - 1$', it is an excellent move to factorize the right hand side (RHS) to obtain '$6(x - 1)/x = (x - 1)(x + 1)$'. It is then tempting to cancel the common factor '$x - 1$' and conclude

$$\text{'} \therefore 6/x = x + 1$$
$$\therefore x^2 + x - 6 = 0, \text{ so } x = 2 \text{ or } x = -3\text{'}.$$

However, the cancellation is valid *only if* $x - 1 \neq 0$, so the above argument is invalid and we have missed one solution. The cancellation must be rewritten as:

$$\text{'}6(x - 1)/x = x^2 - 1 = (x - 1)(x + 1)$$
$$\therefore \text{ either (a) } x - 1 = 0, \text{ or (b) } 6/x = x + 1;$$
$$\therefore \text{ either (a) } \underline{x = 1}, \text{ or (b) } x^2 + x - 6 = 0, \text{ whence } \underline{x = 2 \text{ or } x = -3}\text{ '}.$$

(1.3) Suppose that x and y are any real numbers satisfying $x \cdot y = 0$. If $x \neq 0$, then x has an inverse $1/x$, so we can multiply both sides of $x \cdot y = 0$ to obtain $y = (1/x) \cdot 0$; that is, $y = 0$. This proves the important result that

$$x \cdot y = 0 \implies \text{either } x = 0 \text{ or } y = 0 \,.$$

($*$) Find all real numbers t such that $t(t + 2) = 3$.

(*Hint:* Do not try to solve this directly. Instead, take everything to one side; then factorize to obtain a *product equal to zero*; finally, use (1.3).)

If t can be any real number, one should *never* try to solve $t^2 = at - a$ directly using the factorization $t^2 = a(t - 1)$. This factorization tells you almost nothing about t because a real number can be factorized on infinitely many different ways: for example,

$$2 = 2 \cdot 1 = (2/3) \cdot 3 = \sqrt{2} \cdot \sqrt{2} = (\sqrt{3} - 1)(\sqrt{3} + 1) = \ldots \,.$$

Instead, one must take everything to one side, factorize, and use

$$`x \cdot y = 0 \implies \text{either } x = 0 \text{ or } y = 0'.$$

The warning in the previous paragraph depended on the fact that t could be *any real number*. In contrast, if t and a are *integers*, then a factorization such as $t^2 = a(t - 1)$ tells you a lot because $t - 1$ must then be a factor of t^2! But two consecutive integers, such as $t - 1$ and t, always have a highest common factor equal to 1, so there are only two possible ways in which $t - 1$ can divide t^2, namely, if $t = \underline{\qquad}$ or if $t = \underline{\qquad}$. [Whenever blanks are left in the text, you should make sure you know how to fill them in.]

Another way of finding all integer solutions of the equation $t^2 = at - a$ is to forget for the moment that t and a are *integers*, and to follow the advice 'take everything to one side'. Thus we must find all (integer) solutions of the equation $t^2 - at + a = 0$. If we use the formula for solving quadratic (see the *Algebra* section below), we see that the two real solutions are $t = [a \pm \sqrt{(a^2 - 4a)}]/2$. For $2t$ to be an *integer*, $\sqrt{(a^2 - 4a)}$ must be an integer. Hence $a^2 - 4a$ must be a perfect square. But $a^2 - 4a = (a - 2)^2 - 4$ is also 4 less than a perfect square. Hence the only possible solutions are $a - 2 = 2$ and $a - 2 = -2$; fortunately, in each of these cases the expression $[a \pm \sqrt{(a^2 - 4a)}]$ is not only an integer, but is also even, so the expression for t has an integer value and can be calculated in each case.

The above solution uses *real number* operations carefully and intelligently (avoiding division unless the quotient is an integer, and avoiding square roots unless the number happens to be a perfect square) to solve a problem about *integers*. Sometimes this is impossible, or unnatural, so one must develop special ways of working with integers alone—ways which depend only on addition, subtraction, and multiplication. This leads to *elementary number theory*—the study of *divisibility* (of one integer by another), of *highest common factors* (of pairs of integers), of properties of *prime numbers*, and of *prime factorization*. We will come back to these topics in Section 4.

(1.4) Two natural numbers (in $\mathbb{N}$) can be added or multiplied and the result is always another natural number. In contrast, subtraction and division may give answers outside $\mathbb{N}$. Two integers (in $\mathbb{Z}$) can be added, multiplied, or subtracted and the result is always in $\mathbb{Z}$, but division will usually give an answer outside $\mathbb{Z}$. The set $\mathbb{Q}$ of rational numbers and the set $\mathbb{R}$ of real numbers admit all four operations (although 'dividing by 0' is forbidden).

(1.5) Once a unit of length has been chosen, the real numbers $\mathbb{R}$ can be made to correspond precisely to the set of all points on the *x*-axis. These form a continuum with no gaps. A *real* variable *x* (or *t*, or ...) can vary *continuously*, so we can take *limits*, draw *graphs*, use *calculus*, and so on. The rational numbers $\mathbb{Q}$ correspond to a subset of points on the *x*-axis. This subset of rational points is very crowded, or *dense*, on the *x*-axis.

(∗) Prove that between any two points *a* and *b* $(a \neq b)$ on the *x*-axis there are infinitely many *rational* points.

Nevertheless, the rational numbers leave masses of gaps—even if your eye cannot see them!

(∗) In Section 3 (on '*Proof*') we prove that $\sqrt{2}$ is not a rational number. Use this fact to show that between any two rational numbers *a* and *b* $(a \neq b)$ there are infinitely many irrational numbers.

(1.6) The rational number $1/3$ is perfectly well-defined. However the *decimal* for $1/3$, namely '0.333333...(forever)' (usually written as $0.\dot{3}$) is more complicated in that it only provides us with a *sequence* of better and better approximations to $1/3$:

$$\frac{3}{10}, \frac{33}{100}, \frac{333}{1000}, \frac{3333}{10000}, \ldots$$

No matter how far along this sequence one goes, the terms are never quite equal to $1/3$—though the error at the first stage is less than $1/10$, at the second stage is less than $1/10^2$, and so on. We say that this sequence *converges* to $1/3$, and that the sequence has *limit* equal to $1/3$.

Some convergent sequences of rational numbers have a rational limit: for example, the sequence

$$\frac{2}{1}, \frac{3}{2}, \frac{4}{3}, \frac{5}{4}, \ldots$$

of rational numbers, the *n*th term of which is equal to $(n + 1)/n$, tends to the rational number 1: indeed, $(n + 1)/n = 1 + (1/n)$, so that the *n*th term is just

'1/n' more than 1, and the excess 1/n 'tends to 0' as n gets larger. But *most* convergent sequences of rational numbers have no rational limit. For example, the sequence

$$1, \frac{14}{10}, \frac{141}{100}, \frac{1414}{1000}, \dots$$

of rational numbers, the nth term of which is given by the first n significant figures in the decimal expansion of $\sqrt{2}$, clearly tends to the *irrational* number $\sqrt{2}$.

(∗) Prove that $0.\dot9 = 1$ and that $0.4\dot9 = 1/2$. What simple fractions are $0.\dot0\dot9$, $0.1\dot0\dot9$, and $0.4\dot45\dot5$ equal to? (Prove your assertions.)

(∗∗) Prove that a real number written as a decimal is equal to a rational number if and only if the decimal eventually recurs.

2. Algebra

One of the turning points in the development of modern science occurred around 1600 when Galileo observed that if one wants to understand the 'Book of Nature', then one must learn the language in which that 'Book' is written—namely mathematics. At the time, that language consisted chiefly of number, geometry, and numerical relationships expressed in geometrical form. Since then it has become increasingly clear that the true 'language' of mathematics is *algebra* (a subject which only developed during the seventeenth century).

2.1. Quadratic equations

If a, b, and c are any given real numbers with $a \neq 0$, then the quadratic equation

$$ax^2 + bx + c = 0$$

has solutions

$$x = \frac{-b \pm \sqrt{(b^2 - 4ac)}}{2a}. \tag{2.1}$$

This formula can be used whether or not the quadratic can be easily factorized. The expression $(b^2 - 4ac)$ is called the *discriminant*: if the discriminant is negative, the equation has no real roots; if the discriminant is zero, the equation has one (repeated) root; if the discriminant is positive, the equation has two real roots.

Real roots will in general be irrational. But if the coefficients are integers or rationals, and the discriminant is a perfect square, the quadratic $ax^2 + bx + c$ factorizes nicely and the roots are rational numbers. Conversely, if the coefficients are integers or rationals and you know that one root x is rational, then the discriminant must be a perfect square. (Why?)

(∗) Show that it is impossible to find non-zero integers x and y satisfying

$$x^2 + xy - y^2 = 0.$$

The quadratic formula (2.1) works for arbitrary coefficients a, b, and c (provided that $a \neq 0$). Its derivation is based on the following very useful technique called *completing the square*.

(∗) Fill in the missing coefficients in the following algebraic calculation:

$$ax^2 + bx + c = a(x^2 + \underline{\quad}x + \underline{\quad}) = a((x + \underline{\quad})^2 + \underline{\quad})$$

$$\therefore ax^2 + bx + c = 0 \iff (x + \underline{\quad})^2 + \underline{\quad} = 0 \iff (x + \underline{\quad}) = \pm\sqrt{\underline{\quad}}$$

$$\iff x = -\underline{\quad} \pm \sqrt{\underline{\quad}}.$$

2.2. *Difference of two squares and generalizations*

You will certainly need the familiar identity

$$a^2 - b^2 = (a - b)(a + b). \tag{2.2}$$

It is also important to know that this is the very simplest instance of an infinite family of easy identities.

(∗) Complete the factorization $x^n - 1 \equiv (x - 1)(x^{n-1} + \ldots + 1)$.

(∗) Complete the factorization $x^n - y^n \equiv (x - y)(x^{n-1} + \ldots + y^{n-1})$.

While the difference of two nth powers can always be factorized, the situation for the *sum* of two nth powers is rather different. But do not be misled by the fact that $a^2 + b^2$ does not factorize.

(∗) Complete the factorization $x^3 + 1 \equiv (x + 1)(\ldots)$.

(∗) Complete the factorization $x^3 + y^3 \equiv (x + y)(\ldots)$.

(∗∗) Complete the factorization $x^3 + y^3 + z^3 - 3xyz \equiv (x + y + z)(\ldots)$.

(∗) For which values of n can we factorize $x^n + 1 \equiv (x + 1)(\ldots)$ and $x^n + y^n \equiv (x + y)(\ldots)$?

(∗) Prove that $2^{55} + 1$ is divisible by 33.

(∗) Prove that $1900^{1990} - 1$ is divisible by 1991.

2.3. *Function notation*

One of the most important ideas in mathematics is the idea of a *function*. A function is a rule which tells us how to operate on an 'input' number to obtain an 'output' number. We can write a function in various ways: for example,

$$f : x \mapsto x^2 + 3 \qquad \text{or} \qquad f(x) = x^2 + 3.$$

For this function $f(5) = 28$, because when we substitute 5 in place of x we get the output 28.

You should be prepared to work:

(1) with *polynomial* functions

$$p(x) = a_n x^n + a_{n-1} x^{n-1} + \ldots + a_1 x + a_0,$$

where the coefficients $a_0, a_1, \ldots, a_{n-1}, a_n$ will usually be given integers (but may sometimes be rational numbers, real numbers, complex numbers, or even unknowns);

(2) with *rational* functions

$$r(x) = p(x)/q(x), \qquad \text{where } p(x) \text{ and } q(x) \text{ are polynomials;}$$

(3) and also with the *trig* (trigonometric) functions $\sin x$, $\cos x$, $\tan x$, and so on, with the *exponential* function e^x, and with the *natural logarithm* function $\ln x$.

2.4. *Remainder Theorem for polynomials*

Any polynomial $f(x)$ of degree $n \geqslant 1$ can be divided by $x - a$ to give

$$f(x) = (x - a) \cdot q(x) + r, \tag{2.4}$$

where the *quotient* $q(x)$ is a polynomial of degree $n - 1$ and the remainder r is a constant.

(2.4.1) *The Remainder Theorem.* The Remainder Theorem states that the remainder r in the identity (2.4) is always equal to $f(a)$.

(∗) Prove this.

(2.4.2) *The Factor Theorem.* In particular, a linear expression $x - a$ divides a polynomial $f(x)$ (with no remainder!) precisely when $f(a) = 0$.

(∗) Derive this from (2.4.1).

Example. Suppose that we want to factorize $f(x) \equiv 6x^4 + 5x^3 - 39x^2 + 4x + 12$. If $f(x) = (x - a) \cdot g(x)$ with a an integer (and $g(x)$ a polynomial with integer coefficients), then a must divide the constant term 12 (Why?). Thus it is natural to try values of a which divide 12.

(∗) (a) Calculate $f(1)$, $f(-1)$, $f(2)$, $f(-2)$. Since $f(2) = 0$, we know that $x - 2$ must be a factor.

(b) Complete the factorization $f(x) = (x - 2)(6x^3 \ldots -6)$.

(c) Let $g(x) = 6x^3 + 17x^2 - 5x - 6$. Why do you *not* need to calculate $g(1)$? Or $g(-1)$? Or $g(-2)$? Calculate $g(2)$, $g(3)$, and $g(-3)$.

(d) Factorize $f(x)$ completely as a product of four linear factors.

2.5. *Binomial Theorem*

Given a set of four objects—say, $\{1, 2, 3, 4\}$—there are exactly six different ways in which one can choose two of these four objects (provided that the order in which the two are chosen does not matter); namely, 12, 13, 14, 23, 24, and 34. In general, given a set of n objects—say, $\{1, 2, 3, \ldots, n\}$—let $\binom{n}{r}$ denote the number of ways of choosing exactly r of them (provided that the order in which the r are chosen does not matter). Thus $\binom{4}{2} = 6$.

To obtain a formula for $\binom{n}{r}$ we need one preliminary idea. Given any r objects, we first need to know how many ways there are of putting these objects in order (1st, 2nd, ..., rth). The object to put in the 1st position can be chosen in r different ways; and once the 1st position has been filled, the 2nd position can be filled in $r - 1$ ways. Since there are r ways of filling the 1st position, and for each way of filling the 1st position there are $r - 1$ ways of filling the 2nd position, there are $r \times (r - 1)$ ways of filling the first two positions. Similarly, there are $r \times (r - 1) \times (r - 2)$ ways of filling the first three positions, and $r \times (r - 1) \times (r - 2) \times \ldots \times 3 \times 2 \times 1 = r!$ ways of filling all r positions.

We now count in two different ways the number of ways of choosing r objects *in order* from $\{1, 2, 3, \ldots, n\}$. On the one hand, we could choose the 1st object in n different ways; and for each choice of the 1st object there are $n - 1$ ways of choosing the 2nd object; and for each choice of the first two objects there are $n - 2$ ways of choosing the 3rd object; and so on until, having chosen the first $r - 1$ objects, there remain $n - (r - 1)$ ways of choosing the rth object: there are thus

$$n \times (n - 1) \times (n - 2) \times \ldots \times (n - (r - 1)) = n!/(n - r)! \quad (2.5.1)$$

ways of choosing r objects in order from $\{1, 2, 3, \ldots, n\}$. However, another way of counting the same set could go like this: there are exactly $\binom{n}{r}$ ways of choosing r objects from $\{1, 2, 3, \ldots, n\}$ (provided that the order of choosing does not matter); and for each such choice of r objects, there are $r!$ different ways of putting them in order: there are thus

$$\binom{n}{r} \times r! \qquad (2.5.2)$$

ways of choosing r objects in order from $\{1,2,3,\ldots,n\}$. Equating (2.5.1) and (2.5.2) we obtain

$$\binom{n}{r} = \frac{n!}{(n-r)!\,r!}. \qquad (2.5.3)$$

(*Note:* There is exactly one way of choosing n objects from $\{1,2,3,\ldots,n\}$—you have to choose them all. We thus want $\binom{n}{n} = 1$. This suggests that in the formula

$$\binom{n}{n} = \frac{n!}{(n-n)!\,n!} = \frac{n!}{0!\times n!} = \frac{1}{0!}$$

we must take $0! = 1$.)

(*) (a) Check that $\binom{0}{0} = 1$, $\binom{1}{0} = 1$, and
$\binom{1}{1} = 1$.

(b) Use (2.5.3) to prove that
$$\binom{n}{r} + \binom{n}{r+1} = \binom{n+1}{r+1}.$$

(c) Use (a) and (b) to show that $\binom{n}{r}$ is the rth entry in the nth row of Pascal's triangle (shown on the right).

```
1  ← 0th row
1  1  ← 1st row
1  2  1
1  3  3  1
1  4  6  4  1
1  .  .  .  .  1
.  .  .  .  .  .
```

Binomial Theorem. For any positive integer n,

$$(1+x)^n = \binom{n}{0} + \binom{n}{1}x + \binom{n}{2}x^2 + \binom{n}{3}x^3 + \ldots + \binom{n}{n}x^n. \qquad (2.5.4)$$

Proof. When multiplying out two identical brackets of the form $(1+x)$, we obtain four terms $1\cdot1$, $1\cdot x$, $x\cdot1$, and $x\cdot x$. When multiplying out n such brackets every term will be of the form $x^r\cdot1^{n-r} = x^r$ for some r ($0 \leqslant r \leqslant n$). To obtain a term x^r we have to choose a factor 'x' from r of the n brackets, and a factor '1' from the remaining $n-r$ brackets: the number of such terms is therefore equal to the number of ways of choosing r brackets (for the factor 'x')—that is, $\binom{n}{r}$. Q.E.D.

(∗) (a) Prove $2^n = \binom{n}{0} + \binom{n}{1} + \binom{n}{2} + \binom{n}{3} + \ldots + \binom{n}{n}$.

(b) Prove $\binom{n}{0} + \binom{n}{2} + \binom{n}{4} + \ldots = \binom{n}{1} + \binom{n}{3} + \binom{n}{5} + \ldots$.

Because of (2.5.4), the numbers $\binom{n}{r}$ are called *binomial coefficients*.

2.6. Inequalities

(2.6.1) Suppose that $a > b$ and that k is any real number. Then the following are always true:

(1) $a + k > b + k$;

(2) $ka > kb$ if k is positive, but $ka < kb$ if k is negative;

(3) $a^2 > b^2$ provided that both a and b are positive.

(∗) $2 > -3$, but 2^2 is not greater than $(-3)^2$. Explain! (Thus $x > y$ does not always imply that $x^2 > y^2$.)

(2.6.2) Suppose that $a > b$ and $c > d$. Then we also have:

(4) $a + c > b + d$;

(5) $ac > bd$ provided that the four numbers a, b, c, and d are all *positive*.

(∗) Find real numbers a, b, c, and d such that $a > b$ and $c > d$, but such that $ac < bd$.

(∗) Prove that the square of any real number x is always $\geqslant 0$. Prove that $x^2 = 0$ if and only if $x = 0$.

Many complicated-looking inequalities depend only on this simple fact that

squares cannot be negative.

(∗) Use this fact to prove that, for any real numbers a and b, we have

$$a^2 + b^2 \geqslant 2ab,$$

and that equality $(a^2 + b^2 = 2ab)$ occurs if and only if $a = b$.

(∗) Prove that $(a + b)/2 \geqslant \sqrt{(ab)}$ whenever a and b are positive real numbers.

The LHS in the above inequality '$(a + b)/2$' is the average, or *arithmetic mean* (AM), of the two numbers a and b; the RHS '$\sqrt{(ab)}$' is the *geometric mean* (GM) of a and b. The inequality $(a+b)/2 \geqslant \sqrt{(ab)}$ is often called the 'AM–GM inequality' for two variables a and b.

(∗) Look at your proof of the inequality $(a+b)/2 \geqslant \sqrt{(ab)}$ to discover exactly when *equality* can hold.

(∗) Prove that $a^2+b^2+c^2 \geqslant ab+bc+ca$ for any three real numbers a, b, and c.

(∗∗) Prove the AM–GM inequality for three variables:

$$\frac{a+b+c}{3} \geqslant \sqrt[3]{abc} \text{ whenever } a, b, \text{ and } c > 0.$$

If a, b, $c > 0$, prove that $(a+b+c)/3 = \sqrt[3]{abc}$ if and only if $a = b = c$.

2.7. *Recurrence relations*

The nth term of a sequence is often denoted by u_n. A *recurrence relation* for the sequence (u_n) is a way of defining later terms of the sequence by a formula involving one or more earlier terms. Thus:

(1) $u_1 = 1$, $u_n = u_{n-1} + 3$ defines the sequence $1, 4, 7, 10, 13, \ldots, 3n-2, \ldots$;

(2) $u_1 = 0$, $u_n = 2u_{n-1} + 1$ defines the sequence $0, 1, 3, 7, 15, \ldots, 2^{n-1}-1, \ldots$;

(3) $u_1 = 1$, $u_n = u_{n-1} + n$ defines the sequence

$$1, 3, 6, 10, 15, \ldots, n(n+1)/2, \ldots;$$

(4) $u_1 = 1$, $u_2 = 1$, $u_n = u_{n-1} + u_{n-2}$ defines the sequence $1, 1, 2, 3, 5, 8, \ldots$ of *Fibonacci* numbers;

(5) $u_0 = 1$, $u_n = nu_{n-1}$ defines the sequence $1, 1, 2, 6, 24, 120, \ldots$ of factorials.

(∗∗∗) Let $u_1 = 1$, $u_n = (n-1)u_{n-1} + 1$. For which values of n is u_n divisible by n? When you think you know, try to give a proper proof that your answer is correct.

A sequence of numbers (u_n) is often generated by some *implicit* property, rather than directly from a recurrence relation. In such a case, one step in understanding the sequence (so that we can eventually find a *formula* for the nth term u_n as a function of n) is to find a recurrence relation which the sequence satisfies.

Example. How many regions are created in the plane by 20 straight lines, no two of which are parallel and no three of which pass through the same point?

Solution. Let r_n be the number of regions created by n such lines.

(∗) Check that $r_0 = 1$, $r_1 = 2$, and $r_2 = 4$.

(∗) Prove that $r_n = r_{n-1} + n$.
   (*Hint:* Given n lines, remove one line L to leave $n - 1$ lines, and r_{n-1}
   regions. The line L crosses all the other $n - 1$ lines; so replacing L cuts
   exactly n regions in two.)

(∗) You could now use the recurrence relation to find r_{20}. However, it is
   much better to find a formula for r_n in terms of n (and then put $n = 20$
   to solve the original problem).

(∗∗) (a) Let s_n denote the number of ways of expressing n as a sum of
      positive integers. Thus $s_1 = 1$, $s_2 = 2$, and $s_3 = 4$ (the four ways are
      3, $2 + 1$, $1 + 2$, and $1 + 1 + 1$). Prove that $s_n = s_{n-1} + s_{n-2} + \ldots + s_1 + 1$. Hence calculate s_{10}. Find a formula for s_n in terms of n.
   (b) Suppose that we decide not to distinguish between '$1 + 2$' and
      '$2 + 1$'; Let σ_n denote the number of ways of expressing n as a sum
      of positive integers when the order of the terms does not matter.
      Thus $\sigma_3 = 3$. Calculate σ_n ($1 \leqslant n \leqslant 6$). Find σ_{10}. Express the nth
      term σ_n in terms of earlier terms. Then try to find a formula for σ_n
      in terms of n.

These examples illustrate a general principle in all counting problems.
When trying to find 'How many?' objects of a specified kind there are, the
key idea is to *count some collection in two ways* and then to equate the
answers. This is exactly what a recurrence relation does: for example, s_n is
defined to count the number of ways of writing n as a sum of positive
integers; counting the same set in another way we obtain $s_{n-1} + s_{n-2} + \ldots + s_1 + 1$. Equating the two answers allows us to obtain a very nice formula
for s_n in terms of n. The apparently similar problem for σ_n is much harder.

3. *Proof*

A substantial part of Olympiads (and of all serious mathematics) is concerned
not with obtaining answers, but with the methods you use to obtain those
answers—and in particular with *proofs*. Learning to present reasons and
proofs in a clear and correct manner is a fundamental part of mathematics.
You will not find it easy, but you cannot do mathematics without it. Without
clear proofs mathematics sinks to the level of mere opinion.
   Whenever you want to solve a problem, one way of getting started is to
experiment with special cases and to try to come up with a conjecture which
you believe to be true. But guesses are more often wrong than right. Even

when your guess happens to be correct, if you cannot prove it properly, or cannot be bothered to give a proper proof, there is no reason why other people should take you seriously.

The sequence of rational numbers

$$(2/1)^1, (3/2)^2, (4/3)^3, (5/4)^4, \ldots, ((n+1)/n)^n, \ldots$$

is in some ways 'natural'. Each term seems to be larger than the one before (although this is not easy to prove), and all the terms seem to be less than 3 (which is not easy to prove either). This suggests strongly that as the terms increase, they get closer and closer to some number $\leqslant 3$. But it is not at all easy to decide for sure which real number the sequence tends to. And even if you manage to guess, it is far from easy to prove that your guess is correct.

(∗) (a) What real number do you think the sequence $(2/1)^1, (3/2)^2$, $(4/3)^3, \ldots$ converges to? (You may be able to *guess* on the basis of numerical experiment, even though you are unlikely to be able to prove at this stage that your guess is correct.)

(b) Suppose you were to use a calculator to work out the nth term $u_n = ((n+1)/n)^n$ when $n = 10^{10}$, or $n = 10^{20}$. Why should you not believe the 'answer'? (Try to explain before you try it.)

(∗∗) Prove that the sequence (u_n) is *increasing*—that is, prove that, for each $n > 1$,

$$(n/(n-1))^{n-1} < ((n+1)/n)^n.$$

(∗∗) Prove that every term of the sequence (u_n) is less than 3—that is, prove that, for each $n \geqslant 1$,

$$((n+1)/n)^n < 3.$$

We look briefly at four standard methods of proof that you will certainly need, and one variation which we shall use in the next section.

3.1. *Proof by deduction*

At bottom, all proofs depend on *deduction*. That is, you start out from what you know to be true, and work logically through a sequence of steps until you arrive at the conclusion that you originally wanted to prove.

Example. If x satisfies $|x^2 - 2 + 1| = 1$, prove that $x = 0$ or $x = 2$.

Proof. Suppose that $|x^2 - 2x + 1| = 1$. If we factorize $x^2 - 2x + 1$ the equation becomes $|(x-1)^2| = 1$. Since $(x-1)^2 \geqslant 0$ for all x, $|(x-1)^2| = (x-1)^2$ for all x, so we can drop the modulus signs and the equation to be solved is just $(x-1)^2 = 1$. Subtract 1 from both sides to obtain $x^2 - 2x = 0$. Now factorize and use (1.3): $x(x-2) = 0$; therefore $x = 0$ or $x = 2$.   Q.E.D.

This example has been written out in detail so that it is easy to read. The same proof could be written out using '$\Rightarrow$' or '$\therefore$' in place of words. Although harder for a beginner to read, these are in some ways clearer:

$$\textit{Claim} \quad |x^2 - 2x + 1| = 1 \;\Rightarrow\; x = 0 \text{ or } x = 2.$$

Proof $|x^2 - 2x + 1| = 1 \Rightarrow |(x-1)^2| = 1$	*Proof* $|x^2 - 2x + 1| = 1$
$\Rightarrow (x-1)^2 = 1$	$\therefore (x-1)^2 = |(x-1)^2| = 1$
$\Rightarrow x^2 - 2x = 0$	$\therefore x^2 - 2x = 0$
$\Rightarrow x(x-2) = 0$	$\therefore x(x-2) = 0$
$\Rightarrow x = 0 \text{ or } x = 2$	$\therefore x = 0 \text{ or } x = 2.$ Q.E.D.

To solve a mathematical problem, you may need to calculate some number, to simplify some expression, or to prove some non-obvious fact. Whatever you are trying to do, it is natural (and sensible) while looking for ideas and *while working in rough* to try to make progress both *forwards* from what you know, and *backwards* from what you want to find or prove. This kind of rough work helps to narrow the gap between what is given and what is sought, and often suggests an approach which might help you to get from one to the other. However, when you come to use these ideas in your final solution, you must only use what you know and what you are given: *a deduction must never start from what you are trying to prove.*

3.2. *Distinguishing between a statement and its converse*

One reason why students get confused is that they often fail to distinguish between a given statement (such as 'if $|x^2 - 2x + 1| = 1$, then either $x = 0$ or $x = 2$'), and the *converse* statement (in this case 'if $x = 0$ or $x = 2$, then $|x^2 - 2x + 1| = 1$').

(3.2.1) Usually, when a statement is true, its converse is *false*.

(3.2.2) More importantly, a statement and its converse are logically completely different. So even when a statement and its converse both happen to be true, the proofs of the two statements are likely to be quite different.
    To prove '$|x^2 - 2x + 1| = 1 \;\Rightarrow\; x = 0$ or $x = 2$' it is no good writing something like

'Suppose $x = 0$ or $x = 2$.
Then substituting in $|x^2 - 2x + 1|$ shows that $|x^2 - 2x + 1| = 1$.'

These two lines prove '$x = 0$ or $x = 2 \Rightarrow |x^2 - 2x + 1| = 1$', which is the *converse* of what was wanted. This confusion must be confronted and resolved.

The statement '$|x^2 - 2x + 1| = 1 \Rightarrow x = 0$ or $x = 2$' can be expressed either as a *necessary* condition, or as a *sufficient* condition, depending on whether one is more interested in the *conclusion* '$x = 0$ or $x = 2$', or the *hypothesis* '$|x^2 - 2x + 1| = 1$'. If one is mainly interested in knowing when the conclusion '$x = 0$ or $x = 2$' holds, one might say that

The condition '$|x^2 - 2x + 1| = 1$' is a *sufficient* condition for

$$\text{'}x = 0 \text{ or } x = 2\text{' to hold.}$$

If one is more interested in the hypothesis '$|x^2 - 2x + 1| = 1$', one might say that

The condition '$x = 0$ or $x = 2$' is a *necessary* condition for

$$\text{'}|x^2 - 2x + 1| = 1\text{' to hold.}$$

(3.2.3) Sometimes you are asked to prove an '*if-and-only-if*' statement (such as '$|x^2 - 2x + 1| = 1$ *if and only if* $x = 0$ or $x = 2$'). You then have to prove *two* things:

(1) the '*if*' part (in our example: 'if $x = 0$ or $x = 2$ then $|x^2 - 2x + 1| = 1$');

  and

(2) the '*only if*' part ('if $|x^2 - 2x + 1| = 1$, then either $x = 0$ or $x = 2$').

That is, in the language of the previous paragraph, you have to prove that the condition '$|x^2 - 2x + 1| = 1$' is both *necessary and sufficient* for '$x = 0$ or $x = 2$' to hold.

3.3. *Proof by contradiction*

One way of showing an assertion is true is to show instead that it *cannot possibly be false*.

(3.3.1) $\sqrt{2}$ is not a rational number.

Proof. Suppose that this claim were false. Then $\sqrt{2}$ would have to be a rational number, so that we would have $\sqrt{2} = m/n$ for some integers m and n. By cancelling we can clearly ensure that m and n are not both even. But then

$$\sqrt{2} = m/n \Rightarrow 2 = m^2/n^2 \Rightarrow 2n^2 = m^2 \Rightarrow m^2 \text{ is an even integer.}$$

Now for m^2 to be even, m itself must be even—say $m = 2M$ (because if m were odd, its square m^2 would have to be odd!). Thus $\sqrt{2} = 2M/n$.

$$\sqrt{2} = 2M/n \Rightarrow 2 = 4M^2/n^2 \Rightarrow n^2 = 2M^2 \Rightarrow n^2 \text{ is an even integer.}$$

But then n must also be even. Hence m and n are both even, contrary to our assumption, so we have a contradiction. *Thus the claim cannot possibly be false, and so must be true.*   Q.E.D.

3.4. *Proof by induction*

Often in mathematics we want to prove not just one statement, but infinitely many statements all at once. For example, instead of proving, as we did above, '$\sqrt{2} = 2^{1/2}$ is irrational', we might want to prove the following.

(3.4.1) '$\sqrt[2^n]{2} = 2^{1/2^n}$ is irrational for every integer $n \geqslant 1$'. That is, the (2^n)th root of 2 is irrational for every integer $n \geqslant 1$.

It is important to realize that the words and symbols in (3.4.1) conceal *infinitely many* different claims—one for each value of $n \geqslant 1$. *Proof by induction* is a miraculous way of escaping from the apparently hopeless task of having to prove each claim in turn—a task that could never be completed. To illustrate how this miracle works, we give first a proof of (3.4.1).

Proof of claim (3.4.1): '$2^{1/2^n}$ is irrational' holds for every integer $n \geqslant 1$. For each individual integer n, let $C(n)$ denote the single claim

$$\text{`}2^{1/2^n}\text{' is irrational'}.$$

We prove that $C(n)$ is true <u>for every $n \geqslant 1$</u> in just *two* steps:

(a) $C(1)$ is certainly true (this is precisely (3.3.1)).

(b) Let n be any integer $\geqslant 1$.
   Claim: Suppose that we have *already proved* that $C(n)$ is true for some particular $n \geqslant 1$. Then $C(n + 1)$ must also be true.
   Proof: Suppose $C(n + 1)$ were false. Then $2^{1/2^{(n+1)}}$ is not irrational, so must be rational: $2^{1/2^{(n+1)}} = a/b$ for some integers a and b. But then $2^{1/2^n} = (2^{1/2^{(n+1)}})^2 = a^2/b^2$ would be a rational number, contradicting $C(n)$. Hence $C(n + 1)$ cannot possibly be false, and so must be true.

We now know two things:

(a) $C(1)$ is true, and (b) whenever $C(n)$ is true (for some $n \geqslant 1$),

so is $C(n + 1)$.

Combining (a) and (b) shows that $C(2)$ is true. Feeding this into (b) shows that $C(3)$ must be true—and so on, thus proving that $C(n)$ is true for all $n \geqslant 1$.   Q.E.D.

Step (b) of any induction proof is often *wrongly* understood as 'going from *n* to *n* + 1'. In reality, the only way to prove that C(*n* + 1) is true in step (b) is to somehow *reduce* things in a way which allows us to use the key assumption that C(*n*) is true. This idea (of *reducing* from '*n* + 1' to '*n*') comes out clearly in our second example.

(3.4.2) '$7^{2n+1} + 15^{2n+1}$ is exactly divisible by 22' for every integer $n \geqslant 0$.

Proof. For each individual integer $n \geqslant 0$, let C(*n*) denote the single claim

$$\text{`}7^{2n+1} + 15^{2n+1} \text{ is a multiple of 22'.}$$

(1) Then C(0) says that '$7^1 + 15^1$ is a multiple of 22', which is correct.

(2) Let *n* be any integer $\geqslant 0$:
   Claim: Suppose that we have already proved that C(*n*) is true for some particular value of $n \geqslant 0$. Then C(*n* + 1) must also be true.
   Proof: We have to prove C(*n* + 1); that is, we must show that

$$\text{`}7^{2(n+1)+1} + 15^{2(n+1)+1} \text{ is a multiple of 22'.}$$

The algebra of powers allows us to rewrite this expression:

$$7^{2(n+1)+1} + 15^{2(n+1)+1} = 7^{2n+3} + 15^{2n+3}$$
$$= 7^2(7^{2n+1} + 15^{2n+1}) + (15^2 - 7^2)15^{2n+1}$$
$$= 7^2(7^{2n+1} + 15^{2n+1}) + (15 - 7)(15 + 7)15^{2n+1}.$$

If C(*n*) is true, then the first bracket $(7^{2n+1} + 15^{2n+1})$ is certainly a multiple of 22, so that the first term on the RHS is definitely a multiple of 22. Also, the second term on the RHS has a factor of $(15 + 7) = 22$. Thus the whole RHS is a multiple of 22. Hence the LHS must be a multiple of 22, so that C(*n* + 1) is true.

We now know two things:

(a) C(0) is true, and (b) whenever C(*n*) is true (for some $n \geqslant 0$),

so is C(*n* + 1).

Combining (a) and (b) shows that C(1) must be true. Feeding this into (b) shows that C(2) is true—and so on, thus proving that C(*n*) is true for all $n \geqslant 0$.   Q.E.D.

3.5. *Strong form of proof by induction*

This is a minor, but frequently useful, variation on Section 3.4. Suppose we want to prove that some assertion C(*n*) is true for every $n \geqslant 1$. It is sometimes much easier to use a slightly different (although logically equivalent) form of proof by induction. Suppose we manage to prove both

(a) that C(1) is true, and

(b) that whenever C(1), C(2), ..., C(n) are all true, so is C(n + 1).

Then combining (a) and (b) shows that C(2) is true. Feeding this (and (a)) into (b) shows that C(3) must be true. Feeding all this back into (b) shows that C(4) must be true—and so on, once again proving that C(n) is true for all $n \geqslant 1$. We shall see an example of this in (4.2.1).

4. *Elementary number theory*

Before we launch into a little 'theory', have a go at this number puzzle.

(∗) (a) Take any two-digit number, subtract the sum of the digits, then divide your answer by nine. What do you get? Explain!
　　(b) What happens if you do the same, starting with a three-digit number?

It is tempting to do exactly as instructed: choose a particular two-digit number, do the stated calculation, and write down an answer. But mathematics is about *insight*, not just answers. We often try to convey this by setting problems that have a slightly surprising answer which demands an 'explanation'. However, an answer can only be surprising if you remain alert.

It is easy to miss the first surprising thing in part (a)—namely that the answer (after subtracting the sum of the digits and dividing by nine) always seems to be an *integer*! The second surprise is that the answer is in fact a very special integer.

If in part (a) you take a particular two-digit number (say 73), you might just guess what is going on. You should then realize that 'Explain' means '*prove* that the answer is always equal to the tens digit of the number you started with'. A particular example can sometimes lead to a promising guess; and you can then try to find some way of proving that your guess is correct.

However, this strategy has its limitations. Suppose that you were to try a similar approach in part (b). If you started with a particular number—say, 124—you would land up with the answer 13! This suggests no obvious 'guess'. Mathematical problems cannot usually be solved by *guessing* on the basis of particular examples. Instead, one needs an approach that will reveal the *inner structure* of the problem: in this case (as in many others) this means using *algebra*. If we start with any two-digit starting number 'ab' $= 10a + b$, and then subtract the sum of the digits $a + b$ and divide by 9, we obtain precisely a. Similarly, if we start with any three-digit number 'abc', subtract the sum of the digits $a + b + c$, and then divide by 9, we obtain precisely $11a + b$.

4.1. *Divisibility*

These simple problems illustrate the general principle underlying *elementary number theory*—namely that questions of divisibility can often be resolved by carefully restricting school *algebra* to the set $\mathbb{Z}$ of integers. Because division is usually impossible in $\mathbb{Z}$, the arithmetic and *algebra* of integers is a little different from that for rational numbers or real numbers. The central idea is that of *divisibility*. This, despite its name, depends only on multiplication—not on division.

If m and n are integers, we say that m *divides* n (exactly!) if $n = m \cdot k$ for some integer k. The number m is then called a *factor* of n, n is called a *multiple* of m, and n is said to be *divisible* by m. The equation $n = m \cdot k$ is a *factorization* of n as a product of the two integers m and k. Thus factors come in pairs: m and $k = n/m$.

(∗) Prove that every integer n divides 0. Which two integers divide 1? Which integers does 1 divide?

(∗) Suppose that x and y are integers and that x divides $(x + y)^2$. Prove that x must then divide y^2. Is it true that x always has to divide y? (If so, prove it; if not, give an example to show that it may be false.)

When told that m and n are integers and that m divides n, many students find it hard to resist the temptation to interpret this fact by writing *fractions* such as n/m. This is most unhelpful, for it is hard to remember that this expression which looks like a fraction is in fact an *integer*. The information that one number divides another should *always* be interpreted as an *equation* involving multiplication with integer factors, and avoiding fractions:

'm divides n' means '$n = m \cdot k$ for some integer k'.

When you work with the resulting equations, you then know that *all* the symbols stand for integers. As we shall see, it is then possible to reason in ways that simply don't work if you mix fractions and integers.

(∗) Suppose that x is a positive integer such that x divides $x^2 + 1$. Write this as an equation. Manipulate this equation to obtain a product of two *integers* which is equal to 1. Hence find all possible values for x.

(∗) Suppose that x is an integer (not necessarily positive) such that $x + 3$ divides $x^2 + 5x + 5$. What does this tell you about x?

4.2. *Primes and prime factorization*

A *prime number* is a natural number $n \geqslant 2$ which *cannot* be factorized as a product of smaller natural numbers. The first few prime numbers are 2, 3, 5, 7, 11, 13, 17, and 19. There are just 25 prime numbers less than 100. A natural number $n \geqslant 2$ which is not prime is said to be *composite*.

($*$) Can the difference between a two-digit number and its 'reverse' ever be a prime number?

The importance of prime numbers stems from the fact that every positive integer (other than 1) can be 'broken down' into a product of prime factors.

Example. $\mathbf{360} = 2 \times 180 = 2 \times 2 \times 90 = 2 \times 2 \times 2 \times 45$

$$= 2 \times 2 \times 2 \times 3 \times 15 = 2 \times 2 \times 2 \times 3 \times 3 \times 5.$$

The reason why 1 is not included as a prime number is that—unlike 2, 3, 5, and so on—1 does not help us to break down, or factorize, other integers.

(4.2.1) Every natural number $n \geqslant 2$ can be completely factorized as a product of prime numbers.

Proof. For each $n \geqslant 2$, let $C(n)$ denote the single claim

'n can be completely factorized as a product of prime numbers'.

(1) $C(2)$ is certainly true, since $n = 2$ is a prime number and so is already completely factorized as a 'product' of just one prime number.

(2) Let n be any integer $\geqslant 2$.
   Claim: Suppose we have already proved that $C(2), C(3), \ldots, C(n)$ are *all* true for this particular value of n. Then $C(n + 1)$ must also be true.
   Proof: If $n + 1$ is a prime number, then it is already 'completely factorized as a product of just one prime number', so that $C(n + 1)$ is true. If $n + 1$ is not a prime number, then it can be factorized as a product of two smaller numbers $n + 1 = m \cdot k$. Since $m < n + 1$, we know that $C(m)$ is true; hence m can be completely factorized as a product of prime numbers; say, $m = p_1 p_2 \cdots p_r$. Similarly, $k < n + 1$, so that $C(k)$ is true and k can be completely factorized as a product of prime numbers; say, $k = q_1 q_2 \cdots q_s$. But then $n + 1 = p_1 p_2 \cdots p_r \cdot q_1 q_2 \cdots q_s$ can also be factorized as a product of prime numbers, so $C(n + 1)$ is true.
Hence, by the strong form of proof by induction, $C(n)$ is true for every $n \geqslant 2$.
                                                                        Q.E.D.

(4.2.1) shows that every natural number $n \geqslant 2$ can be factorized in *at least one* way as a product of primes. In Section 4.4 we shall prove that each natural number $n \geqslant 2$ can be factorized in *exactly* one way as a product of primes.

($**$) Prove that the difference between any integer n and any number n' obtained from n by permuting the decimal digits of n is always composite.

4.3. *The division algorithm and highest common factors*

We shall describe the algorithm first, then look at what it does.

Example. Given any two natural numbers, say $p = 481$ and $q = 91$, we first divide the smaller into the larger number to obtain a remainder r:

$$481 = 91 \times 5 + 26.$$

We then do the same all over again, this time with the pair $q = 91$, $r = 26$, and so on:

$$91 = 26 \times 3 + 13,$$
$$26 = 13 \times 2 + 0.$$

Since the remainders get smaller each time, we will always reach a stage at which the last remainder is equal to **0**.

($*$) Why must the remainders get smaller each time?

In the above example, the last equation occurs at the third stage. This equation with remainder $= 0$ shows that **13** divides **26** exactly. If we use this fact in the previous equation, we see that **13** divides the RHS $= 26 \times 3 + 13$ exactly, and so must divide the LHS $= 91$ exactly. But then in the first equation **13** divides the RHS $= 91 \times 5 + 26$ exactly, and so must divide the LHS $= 481$. Hence

13 *is a common factor of* $p = 481$ *and* $q = 91$.

($*$) Let x be any other factor of **481** and **91**. Then the first equation shows that x must divide $481 - 91 \times 5 = 26$ exactly. The second equation then shows that x must divide **13**. Why?

Hence **13** is the *highest common factor* (HCF) of **481** and **91**. We write this as

$$13 = \text{HCF}(481, 91).$$

The miraculous thing about the division algorithm is that it allows you to find the HCF of two natural numbers *without factorizing them*. Moreover, although it is called the 'division' algorithm, you don't actually have to do any division at all. All that is needed when implementing the algorithm is to do *repeated subtraction* and to stop each time when the remainder r is smaller than the 'divisor' q. The division algorithm has many important consequences.

In the above example, the second equation shows that $13 = 91 - 3 \times 26$. If we then substitute for **26** from the first equation, we obtain

$$\text{HCF}(481, 91) = 13 = 91 - 3 \times 26 = 91 - 3 \times (481 - 5 \times 91)$$
$$= 16 \times 91 - 3 \times 481.$$

(∗) Use the division algorithm to find HCF(23 579, 7651). Then substitute backwards to find integers x and y such that

$$\text{HCF}(23\,579, 7651) = x \cdot 23\,579 + y \cdot 7651.$$

(4.3.1) If we apply the division algorithm to any two natural numbers p and q we obtain a sequence of equations. The last non-zero remainder d is the HCF of p and q. Substituting backwards in the sequence of equations then shows how to express $d = \text{HCF}(p, q)$ as a combination of p and q with integer coefficients x and y:

$$d = \text{HCF}(p, q) = x \cdot p + y \cdot q.$$

(∗) Suppose that a, m, and n are integers satisfying the equation $m^2 = an$, and that m and n have highest common factor equal to 1. Prove that $n = 1$ or $n = -1$.

(∗) Use the previous exercise to prove that, if t and a are integers satisfying $t^2 = a(t - 1)$, then $t = 2$ or $t = 0$.

Many problems involving integers x and y (say) are easier to solve once we know that x and y 'have no common factors'; that is, $\text{HCF}(x, y) = 1$. When $\text{HCF}(x, y) = d \neq 1$, it may be possible by writing $x = d \cdot x'$ and $y = d \cdot y'$ to reduce to a similar problem involving x' and y' (with $\text{HCF}(x', y') = 1$). This trick of first removing common factors often makes hard problems easier.

(∗) Suppose that you know that x and y are integers and that xy divides $x^2 + y^2$. Prove that $x = \pm y$.

(4.3.2) Let p be a prime number, and suppose that m and n are integers such that p divides the product $m \cdot n$ exactly. One way in which this could happen is if p divides m exactly. Suppose that p does not divide m exactly. Then $\text{HCF}(p, m) = 1$ (Why?) so that by (4.3.1) we have

$$\text{HCF}(p, m) = 1 = x \cdot p + y \cdot m \text{ for some integers } x \text{ and } y.$$

$$\text{Therefore } n = xpn + ymn.$$

Now p divides the RHS of this last equation exactly (Why?); so p must divide the LHS exactly—that is, p divides n. Hence

if p is a prime *number* and p divides the product mn,

then p divides m, or p divides n.

(∗) The number $p = 4$ divides the product $m \cdot n = 2 \times 6 = 12$, but 4 does *not* divide either of the two factors 2 and 6. Explain!

(4.3.3) If k is an integer and $k^2 = a \cdot b$ for some natural numbers a and b with HCF$(a, b) = 1$, then a and b must both be perfect squares.

(∗∗) Prove this.

> (*Hint:* Denote this assertion by C(k). Check that C(1) is true. Then suppose that C(1), C(2), ..., C(n) are all true, and prove that C($n + 1$) must then be true: let $(n + 1)^2 = a \cdot b$ with HCF$(a, b) = 1$; if $a = 1$, then $a = 1^2$ and $b = (n + 1)^2$ are both perfect squares; if $a \neq 1$, choose a prime p which divides a, show that p^2 divides a, reduce to $((n + 1)/p)^2 = (a/p^2) \cdot b$, and then use the fact that C$((n + 1)/p)$ is true.)

(∗) $k = 6$ is an integer, and $6^2 = 2 \times 18$ is a perfect square, but neither 2 nor 18 are perfect squares. Explain!

(∗∗) Let d and z be integers such that d^2 divides z^2. Prove that d divides z.

Three integers x, y, and z that satisfy the equation $x^2 + y^2 = z^2$ are called a *Pythagorean triple*. A Pythagorean triple x, y, z such that HCF$(x, y) = 1$ is said to be *primitive*.

(∗) Let x, y, z be an unknown Pythagorean triple. Show that HCF$(x, y) =$ HCF$(y, z) =$ HCF(z, x): let this number be d (say). Thus $x = d \cdot x'$, $y = d \cdot y'$, and $z = d \cdot z'$ for some integers x', y', and z'. Show that x', y', z', is a primitive Pythagorean triple; that is, show that HCF$(x', y') =$ HCF$(y', z') =$ HCF$(z', x') = 1$.

(∗) Let p and q be any two integers.
  (a) Show that $x = p^2 - q^2$, $y = 2pq$, $z = p^2 + q^2$ is always a Pythagorean triple.
  (b) Find values of p and q in (a) to produce the Pythagorean triple 3, 4, 5. Can you find values of p and q to produce the Pythagorean triples

$$6, 8, 10; \ 5, 12, 13; \ 9, 12, 15; \ 8, 15, 17; \ 15, 20, 25; \ 15, 36, 39?$$

  (c) What conditions on p and q guarantee that the Pythagorean triple in part (a) is primitive?

Part (b) of the previous exercise should have convinced you that the formulae in part (a) do not produce *all* Pythagorean triples. To find general formulae which will generate *all* Pythagorean triples, one has to go through two stages: first remove common factors, then find a formula for all *primitive* Pythagorean triples. It is not hard to show that, with minor additional restrictions (namely, (i) HCF$(p, q) = 1$ and (ii) one of p and q is odd and the other is even), the formulae in (a) produce all *primitive* Pythagorean triples; hence a general Pythagorean triple always has the form $x = d(p^2 - q^2)$, $y = 2dpq$, $z = d(p^2 + q^2)$.

(**) Suppose that x and y are positive integers and that $2xy$ divides $x + y^2$. Show that x must be a perfect square. What else can you conclude about x?

(*) Suppose that x and y are integers and that xy divides $(x - y)^2$.
  (a) Show that any prime number which divides x must divide y as well.
  (b) Suppose that p is a prime number such that p^3 divides x. Show that p^3 must also divide y.

(*) Find all pairs of positive integers x, y:
  (a) such that $x \cdot y$ divides $x + y$;
  (b) such that $x + y$ divides $x \cdot y$.

4.4. *Uniqueness of prime factorization and counting factors*

We do not distinguish between two 'different' prime factorizations (such as $2 \times 2 \times 3$ and $2 \times 3 \times 2$) if they involve exactly the same prime factors in a different order.

(4.4.1) Every natural number $n \geqslant 2$ can be factorized in *just one way* as a product of prime numbers.

Proof. We have already proved that every natural number can be factorized as a product of prime numbers in *at least one way*. We now have to show that a natural number $n \geqslant 2$ can never be factorized in *more than one way*. Let $C(n)$ denote this assertion for the particular natural number n.

(1) $C(2)$ is true (2 is prime, so it can be factorized in just one way—namely as '2').

(2) Suppose that for some particular value of $n \geqslant 2$ we have already proved that $C(2), C(3), \ldots, C(n)$ are all true. We show that $C(n + 1)$ must then be true.

If $n + 1$ is a prime number, then $C(n + 1)$ is clearly true.
What if $n + 1$ is not a prime number? In that case we *suppose that $n + 1$ can be factorized in two ways and show that they are in fact identical.* Suppose that $n + 1 = p_1 p_2 \cdots p_r = q_1 q_2 \cdots q_s$ are two ways of factorizing $n + 1$ as a product of prime numbers. Then p_1 is a prime number which divides a product $q_1 \cdot (q_2 \cdots q_s)$. Thus p_1 must divide one of the two factors: either p_1 divides q_1 or p_1 divides $(q_2 \cdots q_s)$.
If p_1 divides q_1, then $p_1 = q_1$ (Why?). If p_1 divides $(q_2 \cdots q_s) = q_2 \cdot (q_3 \cdots q_s)$, then p_1 must divide one of the two factors. Continuing in this way we see that, whatever happens, p_1 is equal to q_1, or p_1 is equal to q_2, or p_1 is equal to $q_3, \ldots$, or p_1 is equal to q_s. We may assume that $p_1 = q_2$ and cancel these two primes to give $p_2 p_3 \cdots p_r = q_2 q_3 \cdots q_s = m$ (say). Then, since $C(m)$ is true, the remaining p's $(p_2, \ldots, p_r)$ must be the same as the

remaining q's $(q_2, \ldots, q_s)$, so the two original factorizations were in fact identical.
                                                                                    Q.E.D.

The above result is what makes prime numbers important: every natural number can be broken down *uniquely* as a product of prime numbers. This has many important consequences; we shall content ourselves here with one relatively modest consequence—a simple way of counting how many different factors a number has.

(∗) How many factors does 360 have?
   (*Hint:* $360 = 2^3 \cdot 3^2 \cdot 5$. Each factor of 360 involves a power of 2 (for which there are four choices), a power of 3 (three choices), and a power of 5 (two choices).)

The key to the previous problem is the uniqueness of prime factorization, in that any factor of 360 must be the product of at most three 2's, at most two 3's, and at most one 5. This simple idea is enough to prove the general result.

(4.4.2) If $n = p_1^{a_1} p_2^{a_2} \cdots p_r^{a_r}$, where $p_1, p_2, \ldots, p_r$ are different prime numbers, then n has exactly $(a_1 + 1)(a_2 + 2) \cdots (a_r + 1)$ different factors.

(∗) Which number less than 50 has the most factors? Which number less than 100 has the most factors? Which number less than 1000 has the most factors?

4.5. *Last digits, working with remainders, and congruences*

(∗) My calculator thinks that $(92)^5 = 6\,590\,815\,230$. Why is this answer obviously wrong? What do you think the right answer should be? (Do not use a calculator!)

(∗) My calculator thinks that $(2)^{32} = 4\,294\,967\,300$. What do you think the right answer should be this time?

Some problems about integers can be solved by considering only the last figure, or *units digit*. When we add or multiply two numbers, the units digit of the *answer* can be obtained by simply adding, or multiplying, the units digits of the two starting numbers. The remaining digits have no effect on the units digit of the answer.

(∗) Find the units digit of the answer to each of the following:
   19×93, 19×94, 1993×1994, 19^{93}, 19^{94}, 1993^{1994}, 1994^{1993}.

(∗) Let $N = 1993^{1994} + 1994^{1993}$.
   (a) Show that N is not a perfect square.
   (b) If N were a perfect cube, what would the units digit of its cube root have to be?

   Although some problems about integers can be solved by thinking about units digits, most *cannot* be solved in this way. One reason is that the units digit of a number written in *base 10* depends on the choice of 'base', and so is not a property of the number itself. The number 21 is equal to $2 \times 10 + 1$, and so has units digit '1' in *base 10*; but in *base 9* it would have units digit equal to '3'.

(∗) 1 is a perfect square, but 11, 101, and 1001 are not perfect squares. Can any other number of the form $10000\ldots00001$ ever be a perfect square?

   The previous problem can be solved very simply—but not by looking at the units digit in *base 10*. We must first generalize the underlying *idea*.
   Everyone knows that

$$\text{'}odd \times odd = odd\text{' and that '}odd + odd = even\text{'}.$$

These familiar facts and those of the previous few paragraphs are special cases of a simple, but much more general theory—the theory of *congruences*.

(4.5.1) Let us start by choosing a 'base' n. Then each integer N can be written as a multiple of n plus a remainder r:

$$N = q \cdot n + r, \qquad \text{where } 0 \leqslant r < n.$$

We write this as

$$N \equiv r \,(\text{mod } n)$$

and say that 'N is congruent to r modulo n'. More generally, we write $a \equiv b\,(\text{mod } n)$ whenever n divides $a - b$.

   Examples. $11 \equiv 1\,(\text{mod } 10)$; $78 \equiv 66\,(\text{mod } 12)$; $28 \equiv 13\,(\text{mod } 5)$.

(4.5.2) Let N' be another integer with remainder r' modulo n; that is, $N' = q' \cdot n + r'$. Show that to find the remainders for $N + N'$ and for $N \cdot N'$ modulo n, we only have to look at $r + r'$ and $r \cdot r'$; that is,

$$N + N' \equiv r + r' \,(\text{mod } n) \text{ and } N \cdot N' \equiv r \cdot r' \,(\text{mod } n).$$

(∗) (a) $5.5 = 25 \equiv 5\,(\text{mod } 10)$, so the remainder '5'(mod 10) is congruent to its own square. Which other remainders (mod 10) are congruent to their own squares? Which remainders (mod 10) are congruent to other squares?
   (b) What does part (a) tell you about the units digits of perfect squares (written in base 10)?
   (c) Which one of 5328 and 5329 could be a perfect square? What must it be the square of? (Do not use a calculator!)

(∗) (a) Which remainders (mod 4) are congruent to perfect squares?
  (b) Suppose that x, y, and z are integers satisfying $x^2 + y^2 = z^2$. Could x and y both be odd? (If so, find an example; if not, prove it is impossible.)

(∗) (a) Suppose that $3 - x \equiv 2 \pmod{10}$. What can you say about x?
  (b) Suppose that $3 - 2x \equiv 2 \pmod{10}$. What can you say about x?
  (c) Suppose that $3 - 2x \equiv 1 \pmod{10}$. What can you say about x?

(∗∗) Suppose that $(3 - x) \cdot (x - 7) \equiv 0 \pmod{10}$. What can you say about x?

Congruence '$\equiv \pmod n$' behaves just like the more familiar equality '$=$' for addition and for subtraction. In general, as the previous problem and the next two problems show, you have to be a little careful when using multiplication —and even more careful when using division or cancellation!

(∗) $2 \times 5 \equiv 0 \pmod{10}$, although neither 2 nor 5 is $\equiv 0 \pmod{10}$. Explain!

(∗) $3 \times 8 \equiv 3 \times 4 \pmod{12}$, but we cannot cancel the 3's on each side. Why not?

However, when the 'base' n happens to be a *prime number*, the congruence relation '$\equiv \pmod n$' behaves almost exactly like the more familiar equality '$=$'. Here are two examples.

(∗) Let p be a prime number and let x and y be natural numbers satisfying

$$x \cdot y \equiv 0 \pmod p.$$

Prove that either $x \equiv 0 \pmod p$ or $y \equiv 0 \pmod p$.

(∗) Let p be a prime number and suppose that $a \cdot x \equiv a \cdot y \pmod p$. Prove that

$$\text{either } a \equiv 0 \pmod p, \text{ or } x \equiv y \pmod p.$$

Hence we can cancel a on both sides of such a congruence—provided that a is not congruent to $0 \pmod p$.

(∗) Let p be a prime number. Then $(p - 1)! \not\equiv 0 \pmod p$.
  (*Hint:* Use proof by contradiction. Suppose that p divides $(p - 1)! = (p - 1) \cdot (p - 2)!$. Since p cannot divide the first factor $p - 1$, p must divide the second factor $(p - 2)!$. Now repeat.)

The next fact is often very useful.

(4.5.3) *Fermat's Little Theorem.* Let p be any prime number. Then $a^p \equiv a \pmod p$ for every integer a.

Proof. (a) Suppose that $a \equiv 0 \pmod p$. Then $a^p \equiv 0 \pmod p$, so $a^p \equiv a \pmod p$.
  (b) Suppose that $a \not\equiv 0 \pmod p$. Then none of

$$a \cdot 1, a \cdot 2, a \cdot 3, \ldots, a \cdot (p - 1)$$

is divisible by p (since if p divides $a \cdot i$, p must divide either a or i, both of which are impossible). Moreover, no two of $a \cdot 1, a \cdot 2, \ldots, a \cdot (p-1)$ are congruent (mod p) (since if $a \cdot i \equiv a \cdot j \pmod{p}$, then $i \equiv j \pmod p$)). But there are only $p - 1$ different non-zero remainders (mod p). Thus

$$a \cdot 1, a \cdot 2, \ldots, a \cdot (p-1) \text{ must be congruent to } 1, 2, \ldots, p-1$$

in some order. Hence the product of the first set of numbers must be congruent to the product of the second set of numbers.

$$\therefore (a \cdot 1) \cdot (a \cdot 2) \cdot \ldots \cdot (a \cdot (p-1)) \equiv 1 \cdot 2 \cdot \ldots \cdot (p-1) \pmod p$$

$$\therefore a^{p-1} \cdot (p-1)! \equiv (p-1)! \pmod p$$

$$\therefore a^{p-1} \equiv 1 \pmod p$$

(where we can cancel $(p-1)!$, because it is not $\equiv 0 \pmod p$).   Q.E.D.

Example. $2^{55} + 1$ is exactly divisible by 11.

Proof. $2^{55} = (2^5)^{11} = (32)^{11}$, and $32^{11} = 32 \pmod{11}$.

$$\therefore 2^{55} + 1 \equiv 32^{11} + 1 \equiv 32 + 1 = 33 \equiv 0 \pmod{11}. \quad \text{Q.E.D.}$$

(∗) Give another proof (cf. Section 2.2) that $2^{55} + 1$ is exactly divisible by 33. (Do not use a calculator.)

(∗) Give another proof that $1990^{1990} - 1$ is exactly divisible by 1991.

5. Geometry

In Olympiad problems there is a tendency to assume that students are familiar with the basic geometry of lines, triangles, and circles. You will have met most of the relevant *facts*, but may not have spent enough time on them to become truly familiar with them. Thus you may find it helpful, before working on this section, to revise the basic properties of:

(1) angles in triangles;

(2) angles and parallel lines (alternate angles, supplementary angles, and so on);

(3) circle theorems (angles in the same segment, angles at the centre of the circle, angles in a semi-circle, cyclic quadrilaterals, and so on);

(4) congruence of triangles (side–side–side and side–angle–side criteria);

(5) similarity of triangles.

In particular, if you wish to claim that two triangles are 'congruent' (or 'similar'), the order in which the vertices are listed matters: ' $\triangle A'B'C'$ is congruent to $\triangle ABC$' means ' $A'B' = AB$, $B'C' = BC$, $C'A' = CA$, $\angle A' = \angle A$, and so on'. If you need to work on elementary geometry, you may find one of the general books listed in 'Some books for your bookshelf' helpful.

5.1. *Pythagoras' Theorem and the cosine rule*

The ability to spot similar triangles and to handle the corresponding ratios of sides is absolutely fundamental, and may need a lot of practice. To illustrate, we begin with a proof of Pythagoras' Theorem, using similar triangles.

(5.1.1) *Claim.* If $\triangle ABC$ has a right angle at A, then $BC^2 = AC^2 + AB^2$.

Proof. Let the perpendicular from A to BC hit BC at D. Then $\angle ADC = \angle BAC = 90°$, and $\angle ACD = \angle ACB$, so that $\angle CAD = \angle ABC$. Thus $\triangle DAC$ and $\triangle ABC$ are similar (angles equal in pairs). Hence $AC/BC = DC/AC$. Similarly, $\angle ADB = 90°$ and $\angle ABD = \angle CBA$, so that $\triangle DBA$ and $\triangle ABC$ are similar (angles equal in pairs). Hence $AB/BC = DB/AB$.

$$\therefore \ BC = DC + DB = AC^2/BC + AB^2/BC, \quad \text{so } BC^2 = AC^2 + AB^2. \quad \text{Q.E.D.}$$

In a triangle ABC it is often convenient to call the angle $\angle BAC$ at the point A just 'A', and to denote the length of the side BC opposite angle A by 'a'. Thus $\triangle ABC$ has angles A, B, and C and sides a, b, and c.

(5.1.2) *The cosine rule* (Pythagoras' Theorem for non-right-angled triangles). In any triangle ABC, we have

$$a^2 = b^2 + c^2 - 2bc \cos A$$

(∗) Prove this.

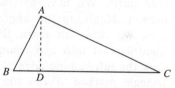

5.2. *Areas of triangles*

We start with the most basic fact of all.

(5.2.1) The area of any triangle is equal to 'half the base × the height'. It follows that:

(1) triangles with equal bases and equal heights have equal areas;

(2) triangles with equal bases have areas proportional to their heights;

(3) triangles with equal heights have areas proportional to their bases;

(4) if $\triangle ABC$ and $\triangle DEF$ are similar, then the ratio of their areas is equal to $(AB/DE)^2$ (and to $(BC/EF)^2$, and to $(CA/FD)^2$).

(5.2.2) Prove that the area of any triangle ABC is equal to $\frac{1}{2}bc \sin A$.

(5.2.3) *Angle bisector theorem.* If the line AX is drawn bisecting the angle A of $\triangle ABC$, then $BX : XC = AB : AC$.

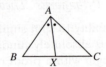

(5.2.4) *Example.* In the triangle on the right the numbers represent the areas of the separate parts. We have to find the unknown area x. Realizing the importance of triangles, we draw an extra line dividing the quadrilateral into two triangles. If we now view the left sloping edge as 'base', then the triangle marked 'a' has the same height as the triangle marked '5', so their bases must be in the ratio $a:5$. Similarly, the triangle $a + b + 8$ has the same height as the triangle

$5 + 10$, so their bases are in the ratio $(a + b + 8):15$. But these two pairs of triangles have the same bases, so that $a/5 = (a + b + 8)/15$, from which we see that $2a - b = 8$.

($*$) Now us the right sloping edge as base, look at the area ratios of two pairs of triangles, and hence show that $b/8 = (a + b + 5)/18$. Hence find x.

5.3. *Intersecting chords theorem*

(5.3.1) If chords AB and CD in a circle meet at X, then $AX \cdot XB = CX \cdot XD$.

Proof.
$\angle BAD = \angle BCD$ (angles in the same segment)

$\angle ADC = \angle ABC$ (angles in the same segment)

$\angle AXD = \angle CXB$ (vertically opposite angles)

$\therefore$ Triangles AXD and CXB are similar.

$\therefore AX : CX = XD : XB$

$\therefore AX \cdot XB = CX \cdot XD$ as required.

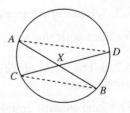

Q.E.D.

(∗) State and prove the corresponding theorem when the chords AB and CD meet *outside* the circle.

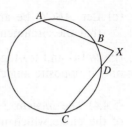

(∗) Suppose that chords CD and AB meet at X outside the circle. As D approaches C, the line CD becomes more and more like the tangent to the circle at C, and $CX \cdot XD$ tends to $(CX)^2$. State and prove the corresponding theorem in this case.

5.4. *Some special points associated with a triangle*

(5.4.1) The *circumcentre* of a triangle (usually labelled O) is the centre of the circle passing through all three vertices. It must therefore be equidistant from all three vertices, and so lies at the intersection of the three perpendicular bisectors of the sides of the triangle.

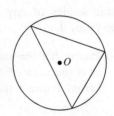

(∗) Prove that if we start with an arbitrary triangle $\triangle ABC$, the three perpendicular bisectors of the sides intersect at the point O.

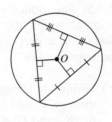

When three or more lines pass through the same point, we say they are *concurrent*. We can therefore rephrase (5.4.1) as follows: 'In any triangle, the perpendicular bisectors of the three sides are concurrent'.

(5.4.2) *The sine rule.* Given any triangle $\triangle ABC$, let its circumcircle have radius R. Draw the diameter BA' through B. Then $\angle BCA' = 90°$ (Why?) and $\angle CA'B = \angle CAB$ (Why?). Thus $\sin A = \sin CA'B = a/BA' = a/2R$. Hence $a/\sin A = 2R$. Similarly, $b/\sin B = 2R$ and $c/\sin C = 2R$. We thus obtain the *sine rule*:

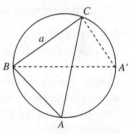

In any triangle $\triangle ABC$, $a/\sin A = b/\sin B = c/\sin C = 2R$.

(5.4.3)
(a) Let $ABCD$ be a cyclic quadrilateral (that is, the four vertices A, B, C, and D lie in order on the circumference of some circle). Show that $\sin A = \sin C$, whence either $A = C$ or $A + C = 180°$; similarly, either $B = D$ or $B + D = 180°$. Show that $A = C$ only if $A = C = 90°$. Hence conclude that we always have $A + C = 180° = B + D$.

(b) Find a more straightforward proof of the fact that opposite angles of a cyclic quadrilateral have sum equal to 180°.

(c) Let *ABCD* be any quadrilateral satisfying $A + C = 180° = B + D$. Show that the circumcircle of $\triangle ABC$ passes through D.

(Parts (a) and (c) together prove that a quadrilateral *ABCD* is cyclic if and only if opposite angles add up to 180°.)

(5.4.4) The *incentre I* of a triangle is the centre of the circle which just touches all three sides. Prove that the incentre lies on the bisector of each angle. Conclude that the bisectors of the three angles of any triangle are concurrent at the point *I*.

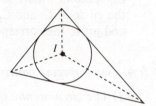

(∗) Prove that the radius *r* of the incircle of $\triangle ABC$ is equal to Δ/s, where Δ is the area of the triangle and $s = (a + b + c)/2$ is half the perimeter. (*Hint:* Express the area Δ in terms of *r*.)

(5.4.5) The *orthocentre H* of a triangle is the point at which the three altitudes meet. (An *altitude* is a perpendicular from one vertex to the opposite side.)

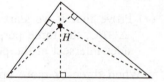

(∗) Given any triangle *ABC*, construct triangle *A′B′C′*, where *A′B′* passes through *C* and is parallel to *AB*, *B′C′* passes through *A* and is parallel to *BC*, and *C′A′* passes through *B* and is parallel to *CA*. Prove that the orthocentre *H* of $\triangle ABC$ is the circumcentre of triangle *A′B′C′*.

(∗) Prove that, if *D*, *E*, and *F* are the feet of the three altitudes of triangle *ABC*, then *H* is the incentre of triangle *DEF*.

(5.4.6) The *centroid G* of a triangle is the point at which the three medians meet. (A *median* is a line joining a vertex to the mid-point of the opposite side.)

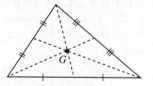

(∗) Prove that in any triangle, the three medians are concurrent.

(∗) Prove that the centroid *G* is two-thirds of the way down each median.

(∗) Prove that any triangle in which $O = I = G = H$ must be equilateral. Does the same conclusion hold if only *some* of the points *O*, *I*, *G*, and *H* coincide?

(∗∗) Prove that in any triangle the points *O*, *G*, and *H* are always collinear (that is, they lie on one straight line, called the *Euler* line) and $OG : GH = 1 : 2$.

If the triangle were cut from a sheet of card, the centroid would be the physical centre of gravity.

6. *Trigonometric formulae*

Our experience of elementary geometry is rooted in the idea that angles are measured in *degrees*. However, the beginnings of higher mathematics (elementary calculus, limits, and so on) reveal that one should really measure angles in *radians*; that is, an angle at C should be measured by the arc-length it determines on the unit circle centred at C. Thus 360° becomes '2π (radians)', 180° becomes 'π (radians)', and so on. In some contexts it does not matter whether one uses degrees or radians; in other contexts important results may only be true if angles are measured in radians. In particular, the trig functions *sin* and *cos* have to be thought of as functions of angles *measured in radians* if one wants to use calculus, or to work with limits. Thus, at the very least, you must learn to be flexible, to think in both degrees and radians, and to sort out when it is essential to work in radians (and why).

You should know by heart the values of

$$\sin 0, \quad \sin \pi/2, \quad \sin \pi, \quad \sin \pi/4, \quad \sin \pi/3, \quad \sin \pi/6;$$
$$\cos 0, \quad \cos \pi/2, \quad \cos \pi, \quad \cos \pi/4, \quad \cos \pi/3, \quad \cos \pi/6.$$

You should also know that 'sin' is an *odd* function, and that 'cos' is an *even* function; that is,

$$\sin(-x) = -\sin(x), \qquad \cos(-x) = \cos(x).$$

You should also memorize (and know how to prove) all the basic trig formulae.

(6.1) $\sin^2 A + \cos^2 A = 1$.

(6.2) $\sin 2A = 2 \sin A \cdot \cos A$.

(6.3) $\cos 2A = 2 \cos^2 A - 1 = 1 - 2 \sin^2 A = \cos^2 A - \sin^2 A$.

(6.4) Use these to derive the following:

(a) $1 + \tan^2 A = \sec^2 A$;

(b) $\sin 2A = 2 \tan A / (1 + \tan^2 A)$;

(c) $\cos 2A = (1 - \tan^2 A)/(1 + \tan^2 A)$;

(d) $\tan 2A = 2 \tan A / (1 - \tan^2 A)$.

(6.5) $\sin(A \pm B) = \sin A \cos B \pm \cos A \sin B$.

(6.6) Use (6.5) to prove that $\sin(\pi/2 - A) = \cos A$. Hence derive the following:

(a) $\cos(A \pm B) = \cos A \cos B \mp \sin A \sin B$;

(b) $\tan(A \pm B) = (\tan A \pm \tan B)/(1 \mp \tan A \tan B)$;

(c) $\sin A \pm \sin B = 2 \sin\left(\dfrac{A \pm B}{2}\right) \cos\left(\dfrac{A \mp B}{2}\right)$;

(d) $\cos A + \cos B = 2 \cos\left(\dfrac{A + B}{2}\right) \cos\left(\dfrac{A - B}{2}\right)$;

(e) $\cos A - \cos B = -2 \sin\left(\dfrac{A + B}{2}\right) \sin\left(\dfrac{A - B}{2}\right)$.

It is important to know quick ways of deriving, or checking, these formulae so that you can always be sure that you are using the correct formula. You must also be ready to use these formulae to prove less familiar results.

You are now ready to have a go at some of the problems in the rest of the book. We all find these problems hard—that is what makes them worth tackling. Good mathematicians do not find maths easy; they simply worry away at every problem until they begin to see what is going on. Half the battle is to learn that things get clearer the longer you keep struggling on. So do resist the temptation to look at the solutions too soon.

Some books for your bookshelf

*If we do not succeed in solving a mathematical
problem, the reason frequently consists in our
failure to recognise the more general standpoint
from which the problem before us appears only
as a single link in a chain of related problems.*
David Hilbert, 1900

The headings used in this section are intended as a rough indication of the contents, and should not be interpreted too strictly. Books which are more demanding are marked with an asterisk (*).

The only useful reference is one that is available. I have therefore restricted myself in the main to books that are (a) in print, or (b) should be available in decent libraries, or (c) may be available in school store cupboards or second-hand bookshops. However, list F gives details of selected resources available from other countries, together with the addresses from which they are available: it would be wise to check prices before ordering, and to include payment with your order.

If you need to learn, or to revise a topic, the first place to look is in the textbooks currently used in your school. If that does not help, you may find that your teachers can help you locate older textbooks which cover the relevant topic in greater depth. School textbooks have the advantage of including lots of exercises to help you come to grips with the ideas. All you need is *one* book that helps you. One book that you really read is better than a whole library of 'useful', but unread, references. The school textbooks listed here are proven texts which may have something to offer, but are no longer widely used—they are not necessarily 'the best'.

More advanced books, not written for a school audience, tend to include fewer worked examples and fewer exercises for you to practise on. You should always read them with a pencil in your hand, and with plenty of rough paper available—actively exploring details as you go along. Learning from books can be difficult, but it is a skill worth mastering.

A. *School textbooks*

S. L. Parsonson (1970/71). *Pure mathematics* (2 vols). Cambridge University Press. [An excellent pure mathematics A-level text, teaching ideas, not rules, and covering everything except calculus; with lots of good (but tough) exercises and answers.]

B. *Problem solving books, each covering a variety of topics*

Ed J. Barbeau, William O. Moser, and Murray S. Klamkin (1995). *Five hundred mathematical challenges*. Mathematical Association of America. [Three hundred pages of problems and solutions to help lay the technical foundations for higher olympiads.]

Judita Cofman (1990). *What to solve? Problems and suggestions for young mathematicians*. Oxford University Press. [An adventure in serious problem solving, built around 120 tough problems—classified by topic and by solution technique.]

Judita Cofman (1995). *Numbers and shapes revisited*. Oxford University Press. [Sequel to 'What to solve?'.]

Heinrich Dörrie (1965). *100 great problems of elementary mathematics*. Dover. [A tour through elementary mathematics guided by classical problems.]

A. Gardiner (1987). *Discovering mathematics*. Oxford University Press. [A hands-on introduction to mathematical 'research' using elementary problems.]

George T. Gilbert, Mark I. Krusemeyer, and Loren C. Larson (1993). *The Wohascum County problem book*. Mathematical Association of America. [One hundred and thirty challenging problems with full solutions.]

Derek Holton, *Problem solving series* (15 booklets). Available from the Mathematical Association (259 London Road, Leicester LE2 3BE). [A chatty introduction to some serious problem solving. Booklets 1, 2, 4, 5, and 6 available as a single book from Canadian Mathematics Competitions (see list F).]

Ross Honsberger (1970). *Ingenuity in mathematics*. Mathematical Association of America.

Ross Honsberger (1973). *Mathematical gems*. Mathematical Association of America.

Ross Honsberger (1976). *Mathematical gems II*. Mathematical Association of America.

Ross Honsberger (1985). *Mathematical gems III*. Mathematical Association of America.

Ross Honsberger (1978). *Mathematical morsels*. Mathematical Association of America.

Ross Honsberger (1979). *Mathematical plums*. Mathematical Association of America.

Ross Honsberger (1985). *More mathematical morsels*. Mathematical Association of America.

Ross Honsberger (1995). *From Erdös to Kiev: problems of Olympiad caliber.* Mathematical Association of America.

Ross Honsberger (1995). *Episodes in nineteenth and twentieth century Euclidean geometry.* Mathematical Association of America. [Honsberger is a beachcomber, who picks up interesting topics as he wanders along the mathematical seashore, and then cuts and polishes them until they sparkle with life—wonderful stuff (includes some exercises and solutions).]

* Loren C. Larson (1983). *Problem-solving through problems.* Springer. [A remarkable array of problems and 'take it or leave it' solutions.]

Hans Rademacher and Otto Toeplitz (1966). *The enjoyment of mathematics.* Princeton. [Twenty-eight readable chapters covering elementary topics from the whole of mathematics, including many neglected areas everyone should meet: now available as a paperback (Dover 1994).]

Hugo Steinhaus (1979). *One hundred problems in elementary mathematics.* Dover. [A classic—elementary problems with genuine mathematical content.]

C. *Books covering essentially one topic*

1. *Numbers*

Ivan Niven (1961). *Numbers rational and irrational.* Mathematical Association of America. [A classic elementary treatment covering all the things one should know (and a bit more); includes exercises and answers.]

David Wells (1986). *The Penguin dictionary of curious and interesting numbers.* Penguin. [Fascinating facts presented in a way which makes you wonder why, and want to know more.]

2. *Proof*

Books on 'proof' are nearly always a disappointment. Proof has to be mastered in context; so the best books on proof are to be found in other parts of this list!

3. *Elementary number theory*

* Alan Baker (1984). *A concise introduction to the theory of numbers.* Cambridge University Press. [If the other books listed here are too easy, this one may help you to see things in a new light!]

Albert H. Beiler (1964). *Recreations in the theory of numbers.* Dover. [Not a textbook—and none the worse for that! Uses serious recreational topics as a springboard for an interesting, though non-systematic, adventure. Lots of problems and some solutions.]

R. P. Burn (1982). *A pathway into number theory*. Cambridge University Press. [Very much a DIY course—800 problems (and very little text) that should let you teach yourself. The hands-on supplement to Davenport's exposition.]

H. Davenport (1992). *The higher arithmetic, an introduction to the theory of numbers* (6th edn). Cambridge University Press (also Dover 1983). [One of the best popular expositions of elementary number theory from A to Z.]

* I. Niven, H. Zuckerman, and H. Montgomery (1991). *An introduction to the theory of numbers* (5th edn). John Wiley. [One of the best, and most complete, number theory textbooks.]

Oystein Ore (1967). *Invitation to number theory*. Mathematical Association of America. [Excellent gentle introduction—with exercises and answers.]

4. *Geometry*

* H. S. M. Coxeter (1961). *Introduction to geometry*. John Wiley. [An amazing panoply of geometry—with exercises and solutions. Inevitably each topic is touched on rather lightly, so be prepared to look elsewhere to fill in the gaps.]

* H. S. M. Coxeter and S. L. Greitzer (1967). *Geometry revisited*. Mathematical Association of America. [Very much a second course. If you know all the geometry in 'A little useful mathematics' and want to use it in less familiar settings, here's your chance (with exercises and solutions).]

C. V. Durell (1921). *A concise geometry*. Bell.

C. V. Durell (1920). *Modern geometry*. Macmillan.

H. G. Forder (1930). *A school geometry*. Cambridge University Press.

H. S. Hall and F. H. Stevens (1903). *A school geometry parts I–VI*. Macmillan. [If your elementary geometry is weak, these old-fashioned texts (reprinted many times) provide a systematic approach with lots of excellent problems.]

H. G. Forder (1931). *Higher course geometry* (2 vols). Cambridge University Press. [Sequel to 'A school geometry'. Lots and lots of excellent problems.]

Harold R. Jacobs (1974). *Geometry*. W. H. Freeman. [A beautifully produced compendium of basic school geometry (and a bit more)—with exercises and selected answers.]

Serge Lang and Gene Morrow (1983). *Geometry: a high school course*. Springer. [A coherent modern treatment of school geometry (with exercises).]

* Z. A. Melzak (1983). *Invitation to geometry*. John Wiley. [A fine exposition of some unusual topics, focused more on problems than on theory.]

D. Pedoe (1995). *Circles: a mathematical view*. Mathematical Association of America.

David Wells (1991). *The Penguin dictionary of curious and interesting geometry*. Penguin. [From Acute-angled triangle dissections to Zonohedra. Every page is bursting with things to do, questions to pursue, results to try to prove.]

5. *Trigonometry*

C. V. Durell and A. Robson (1930). *Advanced trigonometry*. Bell.
H. S. Hall and S. R. Knight (1893). *Elementary trigonometry*. Macmillan.
[Good old school texts (reprinted many times) with lots of problems.]

6. *Algebra*

Ed Barbeau (1989). *Polynomials*. Springer. [A glimpse of higher mathematics
through the study of elementary polynomials. Lots and lots of problems.]
S. Barnard and J. M. Child (1925). *A new algebra*. Macmillan.
H. S. Hall and S. R. Knight (1897). *Algebra for beginners*. Macmillan.
H. S. Hall and S. R. Knight (1885). *Elementary algebra*. Macmillan. [Good old
school texts (reprinted many times) with lots of problems.]
H. S. Hall and S. R. Knight (1887). *Higher algebra*. Macmillan. [Uses algebraic
technique to introduce many related topics from higher mathematics in an
old-fashioned way. Lots of problems.]

7. *Combinatorics (counting)*

V. Boltyansky and I. Gohberg (1985). *Results and problems in combinatorial
geometry*. Cambridge University Press. [A nice introduction to combinatorial geometry (e.g. given a sphere, into how many pieces must one cut it if
all the pieces are to have smaller diameter than the original sphere?). Out
of print, but worth finding.]
Ivan Niven (1965). *Mathematics of choice: how to count without counting*.
Mathematical Association of America. [Excellent introduction to
factorials, binomial coefficients, recurrence relations, inclusion–exclusion,
partitions, induction, etc.]
N. Ya. Vilenkin (1971). *Combinatorics*. Academic Press. [A wonderful collection of counting problems used to motivate the theory (out of print, so try
a good library).]
A. M. Yaglom and I. M. Yaglom (1987). *Challenging mathematical problems
with elementary solutions*, Vol. 1 (*Combinatorial analysis and probability
theory*). Dover. [One hundred excellent (but tough) problems—with the
associated theory discussed in the text; includes full solutions.]

D. *Collections of Olympiad problems, or of mathematical techniques for Olympiad problems*

* Samuel L. Greitzer (1978). *International mathematical olympiads 1959–1977*.
Mathematical Association of America.

* Murray S. Klamkin (1986). *International mathematical olympiads 1978–1985, and forty supplementary problems*. Mathematical Association of America. [These two books cover all the International Olympiads up to 1985. Most of each book is given over to presenting solutions to each problem. Invaluable for the enthusiast, but not much 'motivation' to help beginners. Problems from earlier IMOs tend to be easier than more recent ones.]

Murray S. Klamkin (1988). *USA mathematical olympiads 1972–1986*. Mathematical Association of America. [Similar to the preceding two books—though the problems are simpler than the recent IMOs. The book includes a useful booklist.]

Hungarian problem book I based on the Eötvös competitions 1884–1905. Mathematical Association of America (New Mathematical Library, Vol. 11).

Hungarian problem book II based on the Eötvös competitions 1906–1928. Mathematical Association of America (New Mathematical Library, Vol. 12). [English translations (by Elvira Rapaport) of Joszef Kürschák's 'Problems of the mathematics contest'. Classic problems (with full solutions) from the motherland of mathematical problem solving.]

A. M. Yaglom and I. M. Yaglom (1987). *Challenging mathematical problems with elementary solutions*, Vol. 2. Dover. [Another eighty problems on a variety of topics, together with a brief discussion of the relevant theory; with full solutions.]

D. O. Schlarsky, N. N. Chentzov, and I. M. Yaglom (1962). *The USSR olympiad problem book*. W. H. Freeman (recently reprinted by Dover 1993). [An excellent collection of tough problems. Worth hunting down.]

E. *Puzzle books*

Ed Barbeau (1995). *After Math*. Ward and Emerson Inc. (available from Ward and Emerson, Six O'Connor Drive, Toronto, ON M4K 2K1, Canada). [Lots of lovely puzzles and brain teasers presented with a light touch for the general reader (with solutions and comments on the problems).]

A. Gardiner (1987). *Mathematical puzzling*. Oxford University Press. [Elementary, but genuinely mathematical problems for beginners to cut their teeth on. Reprint available from UK Mathematics Foundation, School of Mathematics, University of Birmingham, B15 2TT, UK.]

Tony Gardiner (1996). *Mathematical challenge*. Cambridge University Press. [Six hundred quality multiple-choice problems to stretch younger secondary pupils.]

Tony Gardiner (1997). *More mathematical challenges*. Cambridge University Press. [More demanding problems for interested students aged 11–15.]

Martin Gardner (1965). *Mathematical puzzles and diversions*. Pelican.

Martin Gardner (1966). *More mathematical puzzles and diversions*. Pelican.

Martin Gardner (1977). *Further mathematical puzzles and diversions*. Pelican.

Martin Gardner (1978). *Mathematical carnival*. Pelican.

Martin Gardner (1981). *Mathematical circus*. Pelican.

Martin Gardner (1987). *Riddles of the sphinx*. Mathematical Association of America.

Martin Gardner (1995). *New mathematical diversions*. Mathematical Association of America. [Real mathematics presented in a chatty way for the serious amateur. Numerous other books by the same author.]

Boris A. Kordemsky (1975). *The Moscow Puzzles*. Pelican. [Hundreds of excellent puzzles (with solutions) involving elementary mathematics; justifiably popular.]

Raymond Smullyan (1981). *What is the name of this book?* Pelican.

Raymond Smullyan (1984). *Alice in puzzleland*. Penguin. [Off-beat, but serious logic problems to sharpen your wits. Numerous other books by the same author.]

Ian Stewart (1989). *Game, set and math*. Penguin. [Real mathematics problems presented in a lively and stimulating way, with lots for readers to get their teeth into. Numerous other books by the same author.]

David Wells (1992). *The Penguin book of curious and interesting puzzles*. Penguin. [Five hundred and sixty-eight classic puzzles (and solutions). Invaluable.]

F. *Books available from non-standard sources*

1. *Available from the Australian Mathematics Trust, University of Canberra, P.O. Box 1, Belconnen ACT 2616, Australia*

J. Edwards, D. King, and P. O'Halloran (eds) (1986). *All the best from the Australian mathematics competition*. Australian Mathematics Competition.

P. O'Halloran, G. Pollard, and P. Taylor (eds) (1992). *More of all the best from the Australian mathematics competition*. Australian Mathematics Competition. [Excellent multiple choice problems for all secondary school ages.]

W. P. Galvin, D. C. Hunt, and P. O'Halloran (eds) (1988). *An olympiad down under*. Australian Mathematics Competition Ltd. [Details of the 1988 IMO in Australia—including all the problems submitted but not used.]

A. Plank and N. Williams (eds) (*c.* 1992—no date). *Mathematical toolchest*. Australian International Centre for Mathematical Enrichment, Canberra. [A slim compendium—including more results than you will ever need, and additional references, but no proofs or exercises.]

J. B. Tabov and P. J. Taylor (1996). *Methods of problem solving* (Book 1). Australian Mathematics Trust. [Lots of problems (and full solutions) on proof by contradiction, induction, rotations, barycentric co-ordinates, and invariants.]

P. J. Taylor (ed.) (1993). *International mathematics Tournament of the Towns 1980–84*. Australian Mathematics Trust.

P. J. Taylor (ed.) (1992). *International mathematics Tournament of the Towns 1984–89*. Australian Mathematics Trust.

P. J. Taylor (ed.) (1994). *International mathematics Tournament of the Towns 1989–93*. Australian Mathematics Trust. [Marvellous tough Russian problems for ages 14–18.]

In addition, the Australian Mathematics Trust (same address) publishes annual reports on its own multiple choice competitions and olympiads, including problems and solutions.

2. *Available from Canadian Mathematical Society, 577 King Edward, Ottawa, Ontario K1N 6N5, Canada*

W. Moser and E. Barbeau (1978). *The first ten Canadian mathematics olympiads 1969–1978*. Canadian Mathematical Society.

C. M. Reis and S. Z. Ditor (1988). *The Canadian mathematics olympiads 1979–1985*. Canadian Mathematical Society. [Problems, solutions, and results from the Canadian olympiads.]

E. Barbeau, M. Klamkin, and W. Moser (1975–80). *1001 problems in high school mathematics* (Books 1–4). Canadian Mathematical Society. [Lots and lots of lovely problems (plus solutions) at sub-olympiad level.]

The Canadian Mathematical Society also publish an excellent problem journal called *Crux Mathematicorum:* see list G.

3. *Available from Canadian Mathematics Competitions, Faculty of Mathematics, University of Waterloo, Waterloo, Ontario N2L 3G1, Canada*

Peter Booth, Bruce Shawyer, and John Grant McLoughlin (1995). *Shaking hands in Corner Brook and other math problems*. Waterloo Mathematics Foundation. [Two hundred problems for senior secondary students (with solutions).]

Canadian Mathematics Competition (1988–96). *Problems, problems, problems*, Vols 1–6. Waterloo Mathematics Foundation. [Multiple choice problems from the Canadian Mathematics Competitions—on all secondary levels.]

D. Holton (1993). *Let's solve some math problems*. Waterloo Mathematics Foundation. [Booklets 1, 2, 4, 5, and 6 from the 'Problem solving series' (see list B), improved and bound together as one.]

4. *Available from MathPro Press, P.O. Box 713, Westford MA 01886-0021, USA*

Stanley Rabinowitz (1992). *Index to mathematical problems*, Vol. 1, *1980–84*. MathPro Press. [Enough problems here for a very long lifetime!]

D. Fomin and A. Kirichenko (1994). *Leningrad mathematical olympiads 1987–1991*. MathPro Press. [Hundreds of excellent problems from one of the most famous olympiads.]

L. Zimmerman and G. Kessler (1995). *ARML–NYSML contests 1989–1994*. MathPro Press. [The UK has nothing to match the American 'math leagues', where teams get together for a couple of days to compete on all sorts of math problems. These are excellent problems that have proved their worth.]

5. *Available from Centre for Excellence in Mathematical Education, 885 Red Mesa Drive, Colorado Springs, CO 80906, USA*

Alex Soifer (1987). *Mathematics as problem solving*. Centre for Excellence in Mathematical Education. [The title says it all. Excellent problems used to suggest something of the essence of mathematics and how to do it better.]

Alex Soifer (1990). *How does one cut a triangle*. Centre for Excellence in Mathematical Education. [Lots of problems (and solutions) involving dissections of triangles.]

V. Boltyanski and A. Soifer (1991). *Geometric etudes in combinatorial mathematics*. Centre for Excellence in Mathematical Education. [Lovely problems from tiling and combinatorial geometry.]

Alex Soifer (1994). *Colorado Mathematical Olympiad*. Centre for Excellence in Mathematical Education. [The first ten Colorado Olympiads together with 'further exploration' of some of the more challenging problems.]

Alex Soifer (1997). *Mathematical colouring book*. Centre for Excellence in Mathematical Education. [Lots of tough and interesting problems involving the combinatorics of colouring.]

6. *Available from Science Culture Technology publishing, AMK Central Post Office, P.O. Box 0581, Singapore 915603*

Marcin Kuczma (1996). *40th Polish mathematics olympiad 1989/1990*. SCT Publishing. [Twenty-four problems—with complete multiple solutions—from the six sets of questions for the 40th Polish olympiad.]

7. *Available from Brendan Kelly Publishing Inc., 2122 Highview Drive, Burlington, Ontario L7R 3X4, Canada*

Ravi Vakil (1996). *A mathematical mosaic: patterns and problem solving.* Brendan Kelly Publishing. [A wonderfully lively and refreshing introduction to selected topics in number theory, combinatorics, geometry, infinity, complex numbers and calculus. Excellent off-beat material for serious students, written by a double IMO gold medallist.]

8. *Available from Deakin University Press, Geelong, Victoria 3217, Australia*

Terence C. S. Tao (1992). *Solving mathematical problems: a personal perspective.* Deakin University Press. [A mature introduction to the gentle art of solving olympiad problems—full of wisdom and good taste, written by another recent IMO gold medallist.]

9. *Available from the Academic Distribution Center, 1216 Walker Road, Freeland, MD 21053, USA*

Marcin E. Kuczma (1994). *Problems: 144 problems of the Austrian–Polish mathematics competition.* The Academic Distribution Centre. [A compilation of the problems posed at the first 16 Austrian–Polish mathematics olympiads 1978–1993, with complete solutions.]

G. *Problem solving journals*

One of the best ways to develop mathematical problem solving is to make a regular habit of tackling challenging problems on an appropriate level. Problem-solving journals, or journals with a regular student 'Problem corner', are a key resource.

The Mathematical Gazette (published by The Mathematical Association, 259 London Road, Leicester, LE2 3BE). [Three substantial volumes each year containing interesting articles and reviews. Each issue contains 'Student problems' and a 'Problem corner' with solutions to previously posed problems.]

Mathematical Spectrum (published by the Applied Probability Trust, Hicks Building, The University, Sheffield S3 7RH). [Three issues each year aimed at interested secondary students, containing articles and a regular problem corner (with solutions).]

Crux Mathematicorum (published by the Canadian Mathematical Society, 577 King Edward, P.O. Box 450, Station A, Ottawa, ON K1N 6N5, Canada). [The problem solver's bible! Eight issues each year, packed with original problems (and solutions) and olympiad problems from around the world.]

Part II
The problems

The problems

The mathematician's strength is refined by fire through solving problems; s/he discovers new methods and fresh ways of looking at things [...] Mathematicians of past centuries devoted themselves with passionate zeal to the solution of specific difficult problems; they appreciated the significance of difficult problems.

David Hilbert, 1900

Background to the problem papers

In the early years up to 1973, the British Mathematical Olympiad consisted of a single round, and selection for the UK International Mathematical Olympiad team each year was based strongly on the results of this one paper. By 1974 it had become clear that one could not select an IMO team on the basis of a single paper. From that year onwards, the BMO was followed by a second three and a half hour paper for 20–40 invited candidates (the Further International Selection Test—or FIST). Thus the BMO became, in effect, a two-stage event. However, both rounds remained strongly *selective*.

Recently, partly as a result of the expansion of other mathematical competitions, the number of candidates wanting to take part in the BMO has increased (the 1995 record entry was around 850, from 300 schools and colleges). Most of these candidates want an additional challenge beyond their school mathematics, but have no experience of the kind of problem usually associated with mathematical 'olympiads'. Thus the BMO has had to adjust. In 1992 the old BMO was re-christened 'BMO, Round 1', and it was implicitly accepted that this would no longer be a mainly *selective* paper, but an event designed to challenge up to a 1000 able, if often technically weak, youngsters in their last years at school.

In the same year (1992), the old FIST became 'BMO, Round 2'—with a purely selective function. (Entry to Round 2 is by invitation only: in recent years the top 50 or so candidates in Round 1 have been invited to enter Round 2 as of right, together with a further 50 or so candidates chosen on the basis of other evidence.)

This book contains the BMO Round 1 papers only. These have generally been taken in January each year. The slight hiatus in the dating of the 1987–9 papers stems from a short-lived attempt to move the date from January to December: the 23rd BMO was held as usual in January 1987 (1987A); the 24th BMO was held in December 1987 instead of January 1988 (and so is dated 1987B); and the 25th BMO was held in December 1988 (dated 1988). The 26th BMO reverted to the more convenient January date in 1990—hence the apparent gap in 1989. Up to 1975, three hours was allowed for each paper; in 1976 the style of problems changed and the time allowed was increased to three and a half hours.

32nd British Mathematical Olympiad, 1996

1. Find as efficiently as possible all pairs (m, n) of positive integers satisfying the following two conditions:
   (a) two of the digits of m are the same as the corresponding digits of n, while the other digits of m are both 1 less than the corresponding digits of n;
   (b) both m and n are four-digit squares.

2. A function f is defined for all positive integers and satisfies

$$f(1) = 1996,$$

and

$$f(1) + f(2) + \ldots + f(n) = n^2 f(n) \qquad \text{for all } n > 1.$$

Calculate the exact value of $f(1996)$.

3. Let ABC be an acute-angled triangle, and let O be its circumcentre. The circle through C, O, and B is called S. The lines AB and AC meet the circle S again at P and Q respectively. Prove that the lines AO and PQ are perpendicular. [*Note:* Given any triangle XYZ, its *circumcentre* is the centre of the circle which passes through the three vertices X, Y, and Z.]

4. For any real number x, let $[x]$ denote the greatest integer which is less than or equal to x. Define $q(n) = \left[\dfrac{n}{[\sqrt{n}]} \right]$ for $n = 1, 2, 3, \ldots$. Determine all positive integers n for which $q(n) > q(n+1)$.

5. Let a, b, and c be positive real numbers. Prove that:
   (a) $4(a^3 + b^3) \geqslant (a + b)^3$;
   (b) $9(a^3 + b^3 + c^3) \geqslant (a + b + c)^3$.

31st British Mathematical Olympiad, 1995

1. Find the first positive integer whose square ends in three 4s. Find all positive integers whose square ends in three 4s. Show that no perfect square ends in four 4s.

2. *ABCDEFGH* is a cube of side 2.
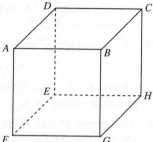
   (a) Find the area of the quadrilateral *AMHN*, where *M* is the midpoint of *BC*, and *N* is the midpoint of *EF*.
   (b) Let *P* be the midpoint of *AB*, and *Q* the midpoint of *HE*. Let *AM* meet *CP* at *X*, and let *HN* meet *FQ* at *Y*. Find the length of *XY*.

3. (a) Find the maximum value of the expression $x^2y - y^2x$ when $0 \leqslant x \leqslant 1$ and $0 \leqslant y \leqslant 1$.
   (b) Find the maximum value of the expression

$$x^2y + y^2z + z^2x - x^2z - y^2x - z^2y$$

   when $0 \leqslant x \leqslant 1$, $0 \leqslant y \leqslant 1$, and $0 \leqslant z \leqslant 1$.

4. Triangle *ABC* has a right angle at *C*. The internal bisectors of angles *BAC* and *ABC* meet *BC* and *CA* at *P* and *Q* respectively. The points *M* and *N* are the feet of the perpendiculars from *P* and *Q* to *AB*. Find angle *MCN*.

5. The seven dwarfs walk to work each morning in single file. As they go they sing their famous song: 'High–low–high–low, it's off to work we go,.... Each day they line up so that no three successive dwarfs are either increasing or decreasing in height. Thus the line-up must go *up–down–up–down–* ... or *down–up–down–up–* .... If they all have different heights, for how many days can they go to work like this if they insist on using a different order each day? What if Snow White always came along too?

30th British Mathematical Olympiad, 1994

1. Starting with any three-digit number n (such as $n = 625$), we obtain a new number $f(n)$ which is equal to the sum of the three digits of n, their three products in pairs, and the product of all three digits.
   (a) Find the value of $n/f(n)$ when $n = 625$. (The answer is an integer!)
   (b) Find all three-digit numbers n such that the ratio $n/f(n) = 1$.

2. In triangle ABC the point X lies on BC.
   (a) Suppose that $\angle BAC = 90°$, that X is the midpoint of BC, and that $\angle BAX$ is one third of $\angle BAC$. What can you say (and prove!) about triangle ACX?
   (b) Suppose that $\angle BAC = 60°$, that X lies one third of the way from B to C, and that AX bisects $\angle BAC$. What can you say (and prove!) about triangle ACX?

3. The sequence of integers $u_0, u_1, u_2, u_3, \ldots$ satisfies $u_0 = 1$ and
$$u_{n+1}u_{n-1} = ku_n \qquad \text{for each } n \geqslant 1,$$
   where k is some fixed positive integer. If $u_{2000} = 2000$, determine all possible values of k.

4. The points Q and R lie on the circle γ, and P is a point such that PQ and PR are tangents to γ. The point A lies on the extension of PQ, and γ' is the circumcircle of triangle PAR. The circle γ' cuts γ again at B, and AR cuts γ at the point C. Prove that $\angle PAR = \angle ABC$.

5. An *increasing* sequence of integers is said to be **alternating** if it *starts* with an *odd* term, the second term is *even*, the third term is *odd*, the fourth is *even*, and so on. The empty sequence (with no terms at all!) is considered to be alternating. Let $A(n)$ denote the number of alternating sequences which only involve integers from the set $\{1, 2, \ldots, n\}$. Show that $A(1) = 2$ and $A(2) = 3$. Find the value of $A(20)$, and prove that your value is correct.

29th British Mathematical Olympiad, 1993

1. Find, showing your method, a six-digit integer n with the following properties:
   (a) the number formed by the last three digits of n is exactly one greater than the number formed by the first three digits of n (so n might look like 123124);
   (b) n is a perfect square.

2. A square piece of toast $ABCD$ of side length 1 and centre O is cut in half to form two equal pieces, ABC and CDA. If the triangle ABC has to be cut into two parts of equal area, one would usually cut along the line of symmetry BO. However, there are other ways of doing this. Find (with proof!) the length and location of the shortest straight cut which divides the triangle ABC into two parts of equal area.

3. For each positive integer c, the sequence u_n of integers is defined by
$$u_1 = 1, \quad u_2 = c, \quad u_n = (2n+1)u_{n-1} - (n^2 - 1)u_{n-2} \qquad (n \geqslant 3).$$
   For which values of c does this sequence have the property that
$$u_i \text{ divides } u_j \text{ whenever } i \leqslant j?$$

[*Note:* If x and y are integers, then x *divides* y if and only if there exists an integer z such that $y = xz$. For example, $x = 4$ divides $y = -12$, since we can take $z = -3$.]

4. Two circles touch internally at M. A straight line touches the inner circle at P and cuts the outer circle at Q and R. Prove that $\angle QMP = \angle RMP$.

5. Let x, y, and z be positive real numbers satisfying

$$\tfrac{1}{3} \leqslant xy + yz + zx \leqslant 3.$$

Determine the range of values for (a) xyz and for (b) $x + y + z$.

28th British Mathematical Olympiad, 1992

1. (a) Observe that the square of 20 has the same number of non-zero digits as the original number. Does there exist a two-digit number *other than* 10, 20, and 30 the square of which has the same number of non-zero digits as the original number? If you think there is one, then give an example. If you claim that there is none, then you must prove your claim.

   (b) Does there exist a three-digit number *other than* 100, 200, and 300 the square of which has the same number of non-zero digits as the original number?

2. Let $ABCDE$ be a pentagon inscribed in a circle. Suppose that AC, BD, CE, DA, and EB are parallel to DE, EA, AB, BC, and CD respectively. Does it follow that the pentagon has to be regular? Justify your claim.

3. Find four distinct positive integers the product of which is divisible by the sum of every pair of them. Can you find a set of five or more numbers with the same property?

4. Determine the smallest value of $x^2 + 5y^2 + 8z^2$, where x, y, and z are real numbers subject to the condition $yz + zx + xy = -1$. Does $x^2 + 5y^2 + 8z^2$ have a *greatest* value subject to the same condition? Justify your claim.

5. Let f be a function mapping positive integers into positive integers. Suppose that

   $$f(n + 1) > f(n) \quad \text{and} \quad f(f(n)) = 3n \qquad \text{for all positive integers } n.$$

   Determine $f(1992)$.

27th British Mathematical Olympiad, 1991

1. Prove that the number $3^n + 2 \times 17^n$, where n is a non-negative integer, is never a perfect square.

2. Find all positive integers k such that the polynomial $x^{2k+1} + x + 1$ is divisible by the polynomial $x^k + x + 1$. For each such k, specify the integers n for which $x^n + x + 1$ is divisible by $x^k + x + 1$.

3. The quadrilateral $ABCD$ is inscribed in a circle of radius r. The diagonals AC and BD meet at E. Prove that if AC is perpendicular to BD, then

$(*)$ $EA^2 + EB^2 + EC^2 + ED^2 = 4r^2.$

Is it true that if $(*)$ holds, then AC is perpendicular to BD? Justify your answer.

4. Find, with proof, the minimum value of $(x + y)(y + z)$, where x, y, and z are positive real numbers satisfying the condition $xyz(x + y + z) = 1$.

5. Find the number of permutations (arrangements)

$$j_1, j_2, j_3, j_4, j_5, j_6 \quad \text{of} \quad 1, 2, 3, 4, 5, 6$$

with the property

for no integer n, $1 \leqslant n \leqslant 5$,

do $j_1, j_2, \ldots, j_n$ form a permutation of $1, 2, \ldots, n$.

6. Show that if x and y are positive integers such that $x^2 + y^2 - x$ is divisible by $2xy$, then x is a perfect square.

7. A ladder of length l rests against a vertical wall. Suppose that there is a rung on the ladder which has the same distance d from both the wall and the (horizontal) ground. Find *explicitly*, in terms of l and d, the height h from the ground that the ladder reaches up the wall.

26th British Mathematical Olympiad, 1990

1. Find a positive integer the first digit of which is 1 and which has the property that, if this digit is transferred to the end of the number, the number is tripled.

2. $ABCD$ is a square and P is a point on the line AB. Find the maximum and minimum values of the ratio PC/PD, showing that these occur for the points P satisfying $AP \times BP = AB^2$.

3. The angles A, B, C, and D of a convex quadrilateral satisfy the relation

$$\cos A + \cos B + \cos C + \cos D = 0.$$

Prove that $ABCD$ is either a trapezium or is cyclic.

4. A coin is biased so that the probability of obtaining a head is p, $0 < p < 1$.

Two players, A and B, throw the coin in turn until one of the sequences HHH or HTH occurs. If the sequence HHH occurs first, then A wins. If HTH occurs first, then B wins. For what value of p is the game fair (that is, such that A and B have an equal chance of winning)?

5. The diagonals of a convex quadrilateral $ABCD$ intersect at O. The centroids of triangles AOD and BOC are P and Q; the orthocentres of triangles AOB and COD are R and S. Prove that PQ is perpendicular to RS. [*Note:* The *centroid* of a triangle is the intersection of the lines joining each vertex to the midpoint of the opposite side; the *orthocentre* is the intersection of the altitudes.]

6. Prove that if x and y are rational numbers satisfying the equation

$$x^5 + y^5 = 2x^2y^2,$$

then $1 - xy$ is the square of a rational number.

25th British Mathematical Olympiad, 1988

1. Find all integers a, b, and c for which

$$(x - a)(x - 10) + 1 = (x + b)(x + c) \qquad \text{for all } x.$$

2. Points P and Q lie on the sides AB and AC respectively of triangle ABC and are distinct from A. The lengths AP and AQ are denoted by x and y repectively, with the convention that $x > 0$ if P is on the same side of A as B, and $x < 0$ on the opposite side; and similarly for y. Show that PQ passes through the centroid of the triangle if and only if

$$3xy = bx + cy,$$

where $b = AC$ and $c = AB$.

3. The lines OA, OB, and OC are mutually perpendicular. Express the area of triangle ABC in terms of the areas of triangles OBC, OCA, and OAB.

4. Consider the triangle of numbers on the right. Each number is the sum of three numbers in the previous row: the number above it and the numbers immediately to the left and right of that number. If there is no number in one or more of these positions, 0 is used. Prove that, from the third row on, every row contains at least one even number.

$$
\begin{array}{ccccccccc}
 & & & & 1 & & & & \\
 & & & 1 & 1 & 1 & & & \\
 & & 1 & 2 & 3 & 2 & 1 & & \\
 & 1 & 3 & 6 & 7 & 6 & 3 & 1 & \\
 & \cdot & \cdot & \cdot & \cdot & \cdot & \cdot & \cdot & \cdot \\
\end{array}
$$

5. None of the angles of a triangle ABC exceeds 90°. Prove that

$$\sin A + \sin B + \sin C > 2.$$

6. The sequence $\{a_n\}$ of integers is defined by $a_1 = 2$, $a_2 = 7$, and

$$-\tfrac{1}{2} < a_{n+1} - \frac{a_n^2}{a_{n-1}} \leqslant \tfrac{1}{2} \qquad \text{for } n \geqslant 2.$$

Prove that a_n is odd for all $n > 1$.

24th British Mathematical Olympiad, 1987B

1. Find all real solutions x of the equation

$$\sqrt{(x + 1\,972\,098 - 1986\sqrt{(x + 986\,049)})}$$

$$+ \sqrt{(x + 1\,974\,085 - 1988\sqrt{(x + 986\,049)})} = 1.$$

   [*Note:* $\sqrt{y}$ always denotes the non-negative square root of y.]

2. Find all real-valued functions f which are defined on the set D of natural numbers $x \geqslant 10$, and which satisfy the functional equation

$$f(x + y) = f(x) \cdot f(y),$$

   for all $x, y \in D$.

3. Find a pair of integers r, s such that $0 < s < 200$ and

$$\frac{45}{61} > \frac{r}{s} > \frac{59}{80}.$$

   Prove that there is exactly one such pair r, s.

4. The triangle ABC has orthocentre H. The feet of the perpendiculars from H to the internal and external bisectors of angle BAC (which is not a right angle) are P and Q. Prove that PQ passes through the midpoint of BC.

5. For any two integers m and n with $0 \leqslant m \leqslant n$, numbers $d(n, m)$ are defined by

$$d(n, 0) = d(n, n) = 1 \qquad \text{for all } n \geqslant 0,$$

   and

$$md(n, m) = md(n - 1, m) + (2n - m)d(n - 1, m - 1) \qquad \text{for } 0 < m < n.$$

   Prove that all of the $d(n, m)$ are integers.

6. Show that, if x and y are real numbers such that $7x^2 + 3xy + 3y^2 = 1$, then the least positive value of $(x^2 + y^2)/y$ is $\tfrac{1}{2}$.

23rd British Mathematical Olympiad, 1987A

1. (a) Find, with proof, all integer solutions of

$$a^3 + b^3 = 9.$$

(b) Find, with proof, all integer solutions of

$$35x^3 + 66x^2y + 42xy^2 + 9y^3 = 9.$$

2. In a triangle ABC, $\angle BAC = 100°$ and $AB = AC$. A point D is chosen on the side AC so that $\angle ABD = \angle CBD$. Prove that $AD + DB = BC$.

3. Find, with proof, the value of the limit as $n \to \infty$ of the quotient

$$\sum_{r=0}^{n} \binom{2n}{2r} \cdot 2^r \bigg/ \sum_{r=0}^{n-1} \binom{2n}{2r+1} \cdot 2^r.$$

[*Note:* $\binom{2n}{s}$ denotes a binomial coefficient.]

4. Let $P(x)$ be any polynomial with integer coefficients such that

$$P(21) = 17, \qquad P(32) = -247, \qquad P(37) = 33.$$

Prove that if $P(N) = N + 51$, for some integer N, then $N = 26$.

5. A line parallel to the side BC of an acute angled triangle ABC cuts the side AB at F and the side AC at E. Prove that the circles on BE and CF as diameters intersect on the altitude of the triangle drawn from A perpendicular to BC.

6. If x, y, and z are positive real numbers, find, with proof, the maximum value of the expression

$$\frac{xyz}{(1+x)(x+y)(y+z)(z+16)}.$$

7. Let n and k be arbitrary positive integers. Prove that there exists a positive integer x such that $\frac{1}{2}x(x+1) - k$ is divisible by 2^n.

22nd British Mathematical Olympiad, 1986

1. Reduce the fraction $\dfrac{N}{D}$ to its lowest terms when

$$N = 2\,244\,851\,485\,148\,514\,627, \qquad D = 8\,118\,811\,881\,188\,118\,000.$$

2. A circle S of radius R has two parallel tangents, t_1 and t_2. A circle S_1 of radius r_1 touches S and t_1; a circle S_2 of radius r_2 touches S and t_2; S_1 also touches S_2. All of the contacts are external. Calculate R in terms of r_1 and r_2.

3. Prove that if m, n, and r are positive integers and

$$1 + m + n\sqrt{3} = (2 + \sqrt{3})^{2r-1},$$

then m is a perfect square.

4. Find, with proof, the largest real number K (independent of a, b, and c) such that the inequality

$$a^2 + b^2 + c^2 > K(a + b + c)^2$$

holds for the lengths a, b, and c of the sides of any *obtuse-angled* triangle.

5. Find, with proof, the number of *permutations*

$$a_1, a_2, a_3, \ldots, a_n \quad \text{of} \quad 1, 2, 3, \ldots, n$$

such that

$$a_r < a_{r+2} \quad \text{for } 1 \leqslant r \leqslant n - 2 \qquad \text{and} \qquad a_r < a_{r+3} \quad \text{for } 1 \leqslant r \leqslant n - 3.$$

[*Note:* A *permutation* is any ordered arrangement of the numbers $1, 2, 3, \ldots, n$.]

6. AB, AC, and AD are three edges of a cube. AC is produced to E so that $AE = 2AC$, and AD is produced to F so that $AF = 3AD$. Prove that the area of the section of the cube by any plane parallel to BCD is equal to the area of the section of tetrahedron $ABEF$ by the same plane.

21st British Mathematical Olympiad, 1985

1. Circles S_1 and S_2 both touch a straight line p at the point P. All points of S_2, except P, are in the interior of S_1. The straight line q: (i) is perpendicular to p; (ii) touches S_2 at R; (iii) cuts p at L; and (iv) cuts S_1 at N and M, where M is between L and R.
   (a) Prove that RP bisects $\angle MPN$.
   (b) If MP bisects $\angle RPL$, find, with proof, the ratio of the areas of the circles S_1 and S_2.

2. Given any three numbers a, b, and c between 0 and 1, prove that not all of the expressions $a(1-b)$, $b(1-c)$, and $c(1-a)$ can be greater than $\frac{1}{4}$.

3. Given any two non-negative integers n and m with $n \geqslant m$, prove that

$$\binom{n}{m} + 2\binom{n-1}{m} + 3\binom{n-2}{m} + \ldots + (n+1-m)\binom{m}{m} = \binom{n+2}{m+2}.$$

[*Note:* $\binom{r}{s}$ denotes the binomial coefficient
$$r(r-1)(r-2)\ldots(r-s+1)/s!.]$$

4. The sequence f_n is defined by $f_0 = 1$ and $f_1 = c$, where c is a positive integer, and for all $n \geqslant 1$ by

$$f_n = 2f_{n-1} - f_{n-2} + 2.$$

Prove that, for each $k \geqslant 0$, there exists h such that $f_k f_{k+1} = f_h$.

5. A circular hoop of radius 4 cm is held fixed in a horizontal plane. A cylinder with radius 4 cm and length 6 cm rests on the hoop with its axis horizontal, and with each of its two circular ends touching the hoop at two points. The cylinder is free to move subject to the condition that each of its circular ends always touches the hoop at two points. Find, with proof, the locus of the centre of one of the cylinder's circular ends.

6. Show that the equation $x^2 + y^2 = z^5 + z$ has infinitely many solutions in positive integers x, y, and z having no common factor greater than 1.

20th British Mathematical Olympiad, 1984

1. P, Q, and R are arbitrary points on the sides BC, CA, and AB respectively of triangle ABC. Prove that the triangle the vertices of which are the centres of the circles AQR, BRP, and CPQ is similar to triangle ABC.

2. Let a_n be the number of binomial coefficients $\binom{n}{r}$ $(0 \leqslant r \leqslant n)$ which leave remainder 1 on division by 3, and let b_n be the number which leave remainder 2. Prove that $a_n > b_n$ for all positive integers n.

3. (a) Prove that, for all positive integers m,

$$\left(2 - \frac{1}{m}\right)\left(2 - \frac{3}{m}\right)\left(2 - \frac{5}{m}\right)\ldots\left(2 - \frac{2m-1}{m}\right) \leqslant m!.$$

(b) Prove that, if a, b, c, d, and e are positive real numbers, then

$$\left(\frac{a}{b}\right)^4 + \left(\frac{b}{c}\right)^4 + \left(\frac{c}{d}\right)^4 + \left(\frac{d}{e}\right)^4 + \left(\frac{e}{a}\right)^4 \geqslant \frac{b}{a} + \frac{c}{b} + \frac{d}{c} + \frac{e}{d} + \frac{a}{e}.$$

4. Let N be a positive integer. Determine, with proof, the number of solutions x in the interval $1 \leqslant x \leqslant N$ of the equation $x^2 - [x^2] = (x - [x])^2$. [*Note:* For a real number x, $[x]$ denotes the *largest integer $\leqslant x$.*]

5. A plane cuts a right circular cone with vertex V in an ellipse E, and meets the axis of the cone at C. The point A is an extremity of the major axis of E. Prove that the area of the curved surface of the slant cone with V as vertex and E as base is $(VA/AC) \times (\text{area of } E)$.

6. Let a and m be positive integers. Prove that if there exists an integer x such that $a^2x - a$ is divisible by m, then there exists an integer y such that both $a^2y - a$ and $ay^2 - y$ are divisible by m.

7. The quadrilateral $ABCD$ has an inscribed circle. To the side AB we associate the expression $u_{AB} = p_1(\sin \angle DAB) + p_2(\sin \angle ABC)$, where p_1 and p_2 are the lengths of the perpendiculars from A and B respectively to the opposite side CD. Define u_{BC}, u_{CD}, and u_{DA} similarly, using in each case the perpendiculars to the opposite side. Show that $u_{AB} = u_{BC} = u_{CD} = u_{DA}$.

19th British Mathematical Olympiad, 1983

1. In triangle ABC with circumcentre O, $AB = AC$, D is the midpoint of AB, and E is the centroid of triangle ACD. Prove that OE is perpendicular to CD.

2. The *Fibonacci sequence* $\{F_n\}$ is defined by

$$F_1 = 1, \quad F_2 = 1, \quad F_n = F_{n-1} + F_{n-2} \quad (n > 2).$$

Prove that there are unique integers a, b, and m such that $0 < a < m$, $0 < b < m$, and $F_n - anb^n$ is divisible by m, for all positive integers n.

3. Given any real number $a \neq -1$, the sequence $x_1, x_2, x_3, \ldots$ is defined by

$$x_1 = a, \quad \text{and} \quad x_{n+1} = x_n^2 + x_n \quad \text{for all } n \geqslant 1.$$

Let $y_n = 1/(1 + x_n)$. Let S_n be the sum, and let P_n be the product, of the first n terms of the sequence $y_1, y_2, y_3, \ldots$. Prove that $aS_n + P_n = 1$, for all n.

4. The two cylindrical surfaces

$$x^2 + z^2 = a^2, \quad z > 0, \quad |y| \leqslant a,$$

and

$$y^2 + z^2 = a^2, \quad z > 0, \quad |x| \leqslant a,$$

intersect. Together with the plane $z = 0$ they enclose a dome-like shape which we shall call a *cupola*. The cupola is placed on top of a vertical tower of height h, the horizontal cross-section of which is a square of side $2a$. Find the shortest distance over the surface of the cupola and tower from the highest point of the cupola to a corner of the base of the tower.

5. Given ten points inside a circle of diameter 5, prove that the distance between some pair of the given points must be less than 2.

6. Consider the equation

$(*)$ $\qquad \sqrt{(2p + 1 - x^2)} + \sqrt{(3x + p + 4)} = \sqrt{(x^2 + 9x + 3p + 9)},$

in which x and p are real numbers and the square roots are to be real (and non-negative). Show that if x and p satisfy $(*)$, then

$$(x^2 + x - p)(x^2 + 8x + 2p + 9) = 0.$$

Hence find all real numbers p for which $(*)$ has just one solution x.

18th British Mathematical Olympiad, 1982

1. The convex quadrilateral $PQRS$ has area A; O is a point inside $PQRS$. Prove that if

$$2A = OP^2 + OQ^2 + OR^2 + OS^2,$$

then $PQRS$ is a square with O as its centre.

2. When written in base 2, a multiple of 17 contains exactly three digits 1. Prove that it contains at least six digits 0, and that if it contains exactly seven digits 0, then it is even.

3. If $s_n = 1 + \frac{1}{2} + \frac{1}{3} + \frac{1}{4} + \ldots + \frac{1}{n}$ and $n > 2$, prove that

$$n(n + 1)^a - n < s_n < n - (n - 1)n^b,$$

where a and b are given in terms of n by $an = 1$ and $b(n - 1) = -1$.

4. For each choice of real number u_1, a sequence $u_1, u_2, u_3, \ldots$ is defined by the recurrence relation $u_n^3 = u_{n-1} + \frac{15}{64}$ ($n \geqslant 2$). By considering the curve $x^3 = y + \frac{15}{64}$, or otherwise, describe, with proof, the behaviour of u_n as n tends to infinity.

5. A right circular cone stands on a horizontal base of radius r. Its vertex V is at a distance l from each point on the perimeter of the base. A plane section of the cone is an ellipse with lowest point L and highest point H.

On the curved surface of the cone, to one side of the plane VLH, two routes, R_1 and R_2, from L to H are marked: R_1 follows the semi-perimeter of the ellipse, while R_2 is the route of shortest length. Find the condition that R_1 and R_2 intersect between L and H.

6. Prove that the number of sequences $a_1, a_2, \ldots, a_n$, with each $a_i = 0$ or 1 and containing exactly m occurrences of '01', is $\binom{n+1}{2m+1}$.

17th British Mathematical Olympiad, 1981

1. The point H is the orthocentre of triangle ABC. The midpoints of BC, CA, and AB are A', B', and C' respectively. A circle with centre H cuts the sides of triangle $A'B'C'$ (produced if necessary) in six points: D_1 and D_2 on $B'C'$; E_1 and E_2 on $C'A'$; and F_1 and F_2 on $A'B'$. Prove that $AD_1 = AD_2 = BE_1 = BE_2 = CF_1 = CF_2$.

2. Given positive integers m and n, S_m is equal to the sum of m terms of the series

$$(n+1) - (n+1)(n+3)$$

$$+ (n+1)(n+2)(n+4) - (n+1)(n+2)(n+3)(n+5) + \ldots,$$

the terms of which alternate in sign, with each term (after the first) equal to the product of consecutive integers with the last but one integer omitted. Prove that S_m is divisible by $m!$, but not necessarily by $m!(n+1)$.

3. Let a, b, and c be any positive numbers. Prove that:
   (a) $a^3 + b^3 + c^3 \geqslant a^2b + b^2c + c^2a$;
   (b) $abc \geqslant (a+b-c)(b+c-a)(c+a-b)$.

4. Given n points in space such that no plane passes through any four of them, let S be the set of all tetrahedra the vertices of which are four of the n points. Prove that a plane which does not pass through any of the n points cannot cut more than $n^2(n-2)^2/64$ of the tetrahedra of S in quadrilateral cross-sections.

5. Find, with proof, the smallest possible value of $|12^m - 5^n|$, where m and n are positive integers.

6. Given distinct non-zero integers a_i $(1 \leqslant i \leqslant n)$, let p_i be the product of all the factors $(a_i - a_1), (a_i - a_2), \ldots, (a_i - a_n)$ except for the zero factor $(a_i - a_i)$. Prove that if k is a non-negative integer, $\sum_{i=1}^{n} a_i^k / p_i$ is an integer.

16th British Mathematical Olympiad, 1980

1. Prove that the equation $x^n + y^n = z^n$, where n is an integer >1, has no solution in integers x, y, and z with $0 < x \leqslant n$ and $0 < y \leqslant n$.

2. Find a set S of seven consecutive positive integers for which a polynomial $P(x)$ of degree 5 exists with the following properties:
   (a) all of the coefficients in $P(x)$ are integers;
   (b) $P(n) = n$ for five numbers $n \in S$, including the least and the greatest;
   (c) $P(n) = 0$ for some $n \in S$.

3. Given a semi-circular region with diameter AB, P and Q are two points on the diameter AB, and R and S are two points on the semi-circular arc such that $PQRS$ is a square. C is a point on the semi-circular arc such that the areas of the triangle ABC and the square $PQRS$ are equal. Prove that a straight line passing through one of the points R and S and through one of the points A and B cuts a side of the square at the incentre of the triangle.

4. Find the set of real numbers a_0 for which the infinite sequence $\{a_n\}$ of real numbers defined by $a_{n+1} = 2^n - 3a_n$ $(n \geqslant 0)$ is strictly increasing. [*Note:* The sequence $\{a_n\}$ is *strictly increasing* if $a_n < a_{n+1}$ for all $n \geqslant 0$.]

5. In a party of ten people, you are told that among any three people there are at least two who do not know each other. Prove that the party contains a set of four people, none of whom knows the other three.

15th British Mathematical Olympiad, 1979

1. Find all triangles ABC for which $AB + AC = 2$ and $AD + BD = \sqrt{5}$, where AD is the altitude through A, meeting BC at the point D.

2. From a point O in three-dimensional space three given rays, OA, OB, and OC, emerge, with $\angle BOC = \alpha$, $\angle COA = \beta$, and $\angle AOB = \gamma$, $0 < \alpha, \beta, \gamma < \pi$. Prove that, given $2s > 0$, there are unique points X, Y, and Z on OA, OB, and OC respectively such that the triangles YOZ, ZOX, and XOY have the same perimeter $2s$. Express OX in terms of s, $\sin(\alpha/2)$, $\sin(\beta/2)$, and $\sin(\gamma/2)$.

3. $S = \{a_1, a_2, \ldots, a_n\}$ is a set of distinct positive odd integers. The differences $|a_i - a_j|$ $(1 \leqslant i < j \leqslant n)$ are all distinct. Prove that

$$\sum_{i=1}^{n} a_i \geqslant \tfrac{1}{3}n(n^2 + 2).$$

4. The function f is defined on the rational numbers and takes only rational values. For all rationals x and y,

$$f(x+f(y))=f(x)f(y).$$

Prove that f is a constant function.

5. If n is a positive integer, denote by $p(n)$ the number of ways of expressing n as the sum of one or more positive integers. Thus $p(4)=5$, as there are five different ways of expressing 4 in terms of positive integers; namely,

$$1+1+1+1,\quad 1+1+2,\quad 1+3,\quad 2+2,\quad \text{and}\quad 4.$$

Prove that $p(n+1)-2p(n)+p(n-1)\geqslant 0$ for each $n>1$.

6. Prove that there are no prime numbers in the infinite sequence

$$10001, 100010001, 1000100010001,\dots .$$

14th British Mathematical Olympiad, 1978

1. Determine, with proof, the point P inside a given triangle ABC for which the product $PL\cdot PM\cdot PN$ is a maximum, where L, M, and N are the feet of the perpendiculars from P to BC, CA, and AB respectively.

2. Prove that there is no proper fraction m/n with denominator $n\leqslant 100$, the decimal expansion of which contains the block of consecutive digits '167', in that order.

3. Show that there is one and only one sequence $\{u_n\}$ of integers such that

$$u_1=1,\quad u_1<u_2,\quad \text{and}\quad u_n^3+1=u_{n-1}u_{n+1}\qquad \text{for all } n>1.$$

4. An *altitude* of a tetrahedron is a straight line through a vertex which is perpendicular to the opposite face. Prove that the four altitudes of a tetrahedron are concurrent if and only if each edge of the tetrahedron is perpendicular to its opposite edge.

5. Inside a cube of side 15 units there are 11 000 given points. Prove that there exists a sphere of unit radius containing at least six of the given points.

6. Show that if n is a non-zero integer, $2\cos n\theta$ is a polynomial of degree n in $2\cos\theta$. Hence or otherwise, prove that if k is rational, then $\cos k\pi$ is either equal to one of the numbers 0, $\pm\frac{1}{2}$, or ± 1, or is irrational.

13th British Mathematical Olympiad, 1977

1. A non-negative integer $f(n)$ is assigned to each positive integer n in such a way that the following conditions are satisfied:
   (a) $f(mn) = f(m) + f(n)$, for all positive integers m, n;
   (b) $f(n) = 0$ whenever the units digit of n (in base 10) is a '3'; and
   (c) $f(10) = 0$.
   Prove that $f(n) = 0$, for all positive integers n.

2. The sides BC, CA, and AB of a triangle touch a circle at X, Y, and Z respectively. Prove that the centre of the circle lies on the straight line through the midpoints of BC and of AX.

3. (a) Prove that if x, y, and z are non-negative real numbers, then
$$x(x-y)(x-z) + y(y-z)(y-x) + z(z-x)(z-y) \geqslant 0.$$
   (b) Hence or otherwise, show that, for all real numbers a, b, and c,
$$a^6 + b^6 + c^6 + 3a^2b^2c^2 \geqslant 2(b^3c^3 + c^3a^3 + a^3b^3).$$

4. The equation $x^3 + qx + r = 0$, where $r \neq 0$, has roots u, v, and w. Express the roots of the equation $r^2x^3 + q^3x + q^3 = 0$ in terms of u, v, and w. Show that if u, v, and w are real, then this latter equation has no root in the interval $-1 < x < 3$.

5. The regular pentagon $A_1A_2A_3A_4A_5$ has sides of length $2a$. For each $i = 1, 2, \ldots, 5$, K_i is the sphere with centre A_i and radius a. The spheres $K_1, K_2, \ldots, K_5$ are all touched externally by each of the spheres P_1 and P_2 —also of radius a. Determine, with proof, whether or not P_1 and P_2 have a common point.

6. The polynomial $26(x + x^2 + x^3 + \ldots + x^n)$, where $n > 1$, is to be decomposed into a sum of polynomials, not necessarily all different. Each of these polynomials is to be of the form $a_1x + a_2x^2 + a_3x^3 + \ldots + a_nx^n$, where each a_i is one of the numbers $1, 2, 3, \ldots, n$ and no two a_i are equal. Find all the values of n for which this decomposition is possible.

12th British Mathematical Olympiad, 1976

1. Find, with proof, the length d of the shortest straight line which bisects the area of an arbitrary given triangle, but which does not pass through any of

the vertices. Express d in terms of the area Δ of the triangle and one of its angles. Show that there is always a shorter line (not straight) which bisects the area of the given triangle.

2. Prove that if x, y, and z are positive real numbers, then

$$\frac{x}{y+z} + \frac{y}{z+x} + \frac{z}{x+y} \geqslant \frac{3}{2}.$$

3. $S_1, S_2, \ldots, S_{50}$ are subsets of a finite set E. Each subset contains more than half the elements of E. Show that it is possible to find a subset F of E having not more than five elements, such that each S_i $(1 \leqslant i \leqslant 50)$ has an element in common with F.

4. Prove that if n is a non-negative integer, then $19 \times 8^n + 17$ is not a prime number.

5. Prove that, if α and β are real numbers, and r and n are positive integers with r odd and $r \leqslant n$, then

$$\sum_{t=0}^{(r-1)/2} \binom{n}{r-t}\binom{n}{t}(\alpha\beta)^t(\alpha^{r-2t} + \beta^{r-2t})$$

$$= \sum_{t=0}^{(r-1)/2} \binom{n}{r-t}\binom{r-t}{t}(\alpha\beta)^t(\alpha+\beta)^{r-2t}.$$

[*Note:* $\binom{m}{s}$ denotes the coefficient of x^s in the expansion of $(1+x)^m$.]

6. A sphere with centre O and radius r is cut by a horizontal plane distance $r/2$ above O in a circle K. The part of the sphere above the plane is removed and replaced by a right circular cone having K as its base and having its vertex V at a distance $2r$ vertically above O. Q is a point on the sphere on the same horizontal level as O. The plane OVQ cuts the circle K in two points, X and Y, of which Y is the further from Q. P is a point of the cone lying on VY, the position of which can be determined by the fact that the shortest path from P to Q over the surfaces of cone and sphere cuts the circle K at an angle of $45°$. Prove that $VP = \sqrt{3} \cdot r/\sqrt{(1 + 1/\sqrt{5})}$. [*Note:* In a spherical triangle ABC, the sides are arcs of great circles, centre O, and the sides are measured by the angles that they subtend at O. You *may* find the following spherical triangle formulae useful: $\sin a/\sin A = \sin b/\sin B = \sin c/\sin C$; $\cos a = \cos b \cos c + \sin b \sin c \cos A$.]

11th British Mathematical Olympiad, 1975

1. Given that x is a positive integer, find (with proof) all solutions of

$$\left[\sqrt[3]{1}\right] + \left[\sqrt[3]{2}\right] + \ldots + \left[\sqrt[3]{(x^3 - 1)}\right] = 400.$$

[*Note:* [z] denotes the largest integer $\leqslant z$.]

2. The first n prime numbers $2, 3, 5, \ldots, p_n$ are partitioned into two disjoint sets A and B. The primes in A are $a_1, a_2, \ldots, a_h$, and the primes in B are $b_1, b_2, \ldots, b_k$, where $h + k = n$. The two products

$$\prod_{i=1}^{h} a_i^{\alpha_i} \quad \text{and} \quad \prod_{i=1}^{k} b_i^{\beta_i}$$

are formed, where the α_i and β_i are any positive integers. If d divides the difference of these two products, prove that either $d = 1$ or $d > p_n$.

3. Use the pigeonhole principle to solve the following problem. Given a point O in the plane, the disc S with centre O and radius 1 is defined as the set of all points P in the plane such that $|OP| \leqslant 1$, where OP is the distance of P from O. Prove that if S contains seven points such that the distance from any one of the seven points to any other is $\geqslant 1$, then one of the seven points must be at O. [*Note:* The *pigeonhole principle* states that if more than n objects are put into n pigeonholes, some pigeonhole must contain more than one object.]

4. In a triangle ABC, three parallel lines AD, BE, and CF are drawn, meeting the sides BC, CA, and AB in D, E, and F respectively. The points P, Q, and R are collinear and divide AD, BE, and CF respectively in the same ratio $k : 1$. Find the value of k.

5. Let m be a fixed positive integer. You are given that

$$\binom{2m}{0} + \binom{2m}{1}\cos\theta + \binom{2m}{2}\cos 2\theta + \ldots + \binom{2m}{2m}\cos 2m\theta$$

$$= (2\cos\tfrac{1}{2}\theta)^{2m}\cos m\theta,$$

where there are $2m + 1$ terms on the LHS; the value of each side of this identity is defined to be $f_m(\theta)$. The function $g_m(\theta)$ is defined by

$$g_m(\theta) = \binom{2m}{0} + \binom{2m}{2}\cos 2\theta + \binom{2m}{4}\cos 4\theta + \ldots + \binom{2m}{2m}\cos 2m\theta.$$

Given that there is no rational k for which $\alpha = k\pi$, find the values of α for which $\lim_{m \to \infty}[g_m(\alpha)/f_m(\alpha)] = \tfrac{1}{2}$.

6. Prove that, if n is a positive integer greater than 1, and $x > y > 1$, then

$$\frac{x^{n+1} - 1}{x(x^{n-1} - 1)} > \frac{y^{n+1} - 1}{y(y^{n-1} - 1)}.$$

7. Prove that there is only one set of real numbers $x_1, x_2, \ldots, x_n$ such that

$$(1 - x_1)^2 + (x_1 - x_2)^2 + \ldots + (x_{n-1} - x_n)^2 + x_n^2 = \frac{1}{n + 1}.$$

8. The interior of a wine glass is a right circular cone. The glass is half filled with water and is then slowly tilted so that the water reaches a point P on the rim. If the glass is further tilted (so that water spills out), what fraction of the conical interior is occupied by water when the horizontal plane of the water level bisects the generator of the cone furthest from P?

10th British Mathematical Olympiad, 1974

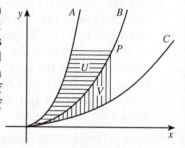

1. The curves A, B, and C are related in such a way that B 'bisects' the area between A and C; that is, the area of the region U is equal to the area of the region V at all points P of the curve B. Find the equation of the curve B given that the equation of curve A is $y = \frac{4}{3}x^2$ and that the equation of curve C is $y = \frac{3}{8}x^2$.

2. A domino can be represented by an unordered pair of integers. Thus ⟨domino⟩ can be represented as $(1, 5)$ or $(5, 1)$ and the double ⟨domino⟩ as $(2, 2)$. The set of all 15 dominoes containing two integers from 1, 2, 3, 4, and 5 is partitioned into three subsets of five dominoes. The dominoes in each subset form a closed chain; that is, $(a, b)(b, c)(c, d)(d, e)(e, a)$, where a, b, c, d, and e need not all be different. How many distinct partitions are there? (The order of the three subsets in the partition is immaterial.)

3. Prove that it is impossible for all the faces of a convex polyhedron to be hexagons.

4. M is a 16×16 matrix. Each entry in the leading diagonal and each entry in the bottom row (that is, the 16th row) is 1. Every other entry of the matrix is $\frac{1}{2}$. Find the inverse of M.

5. A bridge deal is defined as the distribution of 52 ordinary playing cards among four players, so that each player has 13 cards. In a bridge deal,

what is the probability that just one player has a complete suit? (Leave your answer in factorials.)

6. The points X and Y are the feet of the perpendiculars from P to CA and CB respectively, where P is in the plane of triangle ABC. If $PX = PY$, and the straight line through P which is perpendicular to AB cuts XY at Z, prove that CZ bisects AB.

7. The roots of the equation $x^3 = bx + c$ $(bc \neq 0,\ b$ and c real$)$ are α, β, and γ. If

$$\beta = p\alpha^2 + q\alpha + r, \qquad \gamma = p\beta^2 + q\beta + r, \qquad \alpha = p\gamma^2 + q\gamma + r,$$

determine p, q, and r in terms of b and c. State a condition which ensures that p, q, and r are real.

8. Let n be an odd prime number. It is required to write the product

$$\prod_{i=1}^{n-1} (x+i)$$

as a polynomial

$$\sum_{j=0}^{n-1} a_j x^j.$$

By considering the product $\prod_{i=1}^{n}(x+i)$ in two ways, establish the relations

$$a_{n-1} = 1,$$
$$a_{n-2} = n(n-1)/2!,$$
$$2a_{n-3} = n(n-1)(n-2)/3! + a_{n-2}(n-1)(n-2)/2!,$$
$$\ldots$$
$$\ldots$$
$$(n-2)a_1 = n + a_{n-2}(n-1) + a_{n-3}(n-2) + \ldots + 3a_2,$$
$$(n-1)a_0 = 1 + a_{n-2} + \ldots + a_1.$$

Prove that $n \mid a_j$ $(j = 1, 2, \ldots, n-2)$ and that $n \mid (a_0 + 1)$. Prove also that, when x is an integer,

$$n \mid (x+1)(x+2)\ldots(x+n-1) - x^{n-1} + 1.$$

Hence deduce Wilson's Theorem and Fermat's Theorem; namely, that when n is prime and x is not a multiple of n:
(a) $n \mid (n-1)! + 1$;
(b) $n \mid x^{n-1} - 1$.
[*Note:* $p \mid q$ means p divides q leaving no remainder.]

9. A vertical uniform rod of length $2a$ is hinged at its lower end to a frictionless joint secured to a horizontal table. It falls from rest in this unstable position on to the table. Find the time occupied in falling. Comment on your answer. [*Note:* You may quote the result $\int (\operatorname{cosec} x)\mathrm{d}x = \log |(\tan \frac{1}{2}x)|$ if you wish.]

10. A right circular cone the vertex of which is V and the semi-vertical angle of which is α has height h and uniform density. All points of the cone the distances of which from V are less than a or greater than b, where $0 < a < b < h$, are removed. A solid of mass M is left. Given that the gravitational attraction that a point mass m at P exerts on a unit mass at O is $(Gm/OP^3)\vec{OP}$, prove that the magnitude of the gravitational attraction of this solid on a unit mass at V is

$$\tfrac{3}{2}GM(1 + \cos \alpha)/(a^2 + ab + b^2).$$

9th British Mathematical Olympiad, 1973

1. (a) Two fixed circles are touched by a variable circle at P and Q. Prove that PQ passes through one of two fixed points.
   (b) State a true theorem about ellipses (or if you like about conics in general) of which the result in part (a) is a particular case.

2. Nine points are given in the interior of the unit square. Prove that there exists a triangle of area $\leqslant \frac{1}{8}$ the vertices of which are three of the points.

3. A curve consisting of the quarter circle $x^2 + y^2 = r^2$, $x, y \geqslant 0$, together with the line segment $x = r$, $-h \leqslant y \leqslant 0$, is rotated about $x = 0$ to form a surface of revolution which is a hemisphere on a cylinder. A string is stretched tightly over the surface from the point on the curve $(r \sin \theta, r \cos \theta)$ to the point $(-r, -h)$ in the plane of the curve. Show that the string does not lie in a plane if $\tan \theta > r/h$. [*Note:* You may assume spherical triangle formulae such as $\cos a = \cos b \cos c + \sin b \sin c \cos A$, or $\sin A \cot B = \sin c \cot b - \cos c \cos A$. In a spherical triangle the sides a, b, and c are arcs of great circles and are measured by the angles that they subtend at the centre of the sphere.]

4. You have a large number of congruent equilateral triangles on a table and you want to fit n of these together to make a convex equiangular hexagon (that is, one the interior angles of which are each 120°). Obviously, n cannot be *any* positive integer. The smallest n is 6, the next smallest is 10, and the next 13. Determine conditions for possible n.

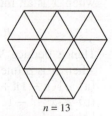

$n = 13$

5. There is an infinite set of positive integers of the form $2^n - 3$ with the following property Q:

   no two members of the set have a common prime factor.

   Here is an outline of a proof of this fact.
   Suppose that there is a finite set $S = \{2^{m_1} - 3, 2^{m_2} - 3, \ldots, 2^{m_k} - 3\}$ with property Q having k members. Let the prime factors of these k numbers be $p_1, p_2, \ldots, p_t$. Consider the number $N = 2^{(p_1-1)(p_2-1)\ldots(p_t-1)+1}$. By Fermat's Theorem, $a^{p-1} \equiv 1 \pmod{p}$ for every prime p that does not divide a. Hence $N - 3 \equiv -1 \pmod{p_r}$, for each r $(1 \leqslant r \leqslant t)$, and $N - 3$ may be added to S to give a larger set with property Q.
   Give a properly expanded and reasoned proof that there is an infinite set of positive integers of the form $2^n - 7$ with property Q.

6. In answering general knowledge questions (framed so that each question is answered 'yes' or 'no') the teacher's probability of being correct is α and a pupil's probability of being correct is β or γ according as the pupil is a boy or a girl. The probability of a randomly chosen pupil agreeing with the teacher's answer is $\frac{1}{2}$. Find the ratio of the number of boys to girls in the class.

7. The life-table issued by the Registrar-General of Draconia shows, out of each $10\,000$ live births, the number y expected to be alive x years later. When $x = 60$, $y = 4820$. When $x = 80$, $y = 3205$. At age 100, all Draconians are put to death. For $60 \leqslant x \leqslant 100$ the curve $y = Ax(100 - x) + B/(x - 40)^2$ fits the figures in the table very closely, A and B being constants. Determine the life-expectancy (in years correct to one decimal place) of a Draconian aged 70.

8. Call the matrix $M = \begin{pmatrix} a & b \\ c & d \end{pmatrix}$ the *companion* matrix for the mapping
   $$T : z \to \frac{az + b}{cz + d}.$$
   (a) Prove that, if M_i is the companion matrix for T_i $(i = 1, 2)$, then $M_1 M_2$ is the companion matrix for the mapping $T_1 T_2$.
   (b) Find conditions on a, b, c, and d so that $T^4 = I$ but $T^2 \neq I$.

9. Let L_r equal the determinant
   $$\begin{vmatrix} x & y & 1 \\ a + c \cos \theta_r & b + c \sin \theta_r & 1 \\ l + n \cos \theta_r & m + n \sin \theta_r & 1 \end{vmatrix}.$$

   Show that the lines $L_r = 0$, $r = 1, 2, 3$, are concurrent and find the co-ordinates of the point at which they meet.

10. Construct a detailed flowchart for a computer program to print out all positive integers up to 100 of the form $a^2 - b^2 - c^2$, where a, b, and c are positive integers and $a \geqslant b + c$. (There is no need to print in ascending order or to avoid repetitions.)

11. (a) Two uniform rough right circular cylinders, A and B, with the same length, have radii a and b and masses M and m respectively. Cylinder A rests with a generator in contact with a rough horizontal table. Cylinder B rests on A, initially in unstable equilibrium, with its axis vertically above that of A. Equilibrium is disturbed, B rolls on A, and A rolls on the table. In the subsequent motion, the plane containing the axes makes an angle θ with the vertical. Draw diagrams showing angles, forces, etc., for the period during which there is no slipping. Write down equations which will give on elimination a differential equation for θ, stating the principles used. *Indicate* how the elimination could be done (you are not asked to *do* it).

    (b) Such a differential equation is, with $k = M/m$,

$$\ddot{\theta}(4 + 2\cos\theta - 2\cos^2\theta + 9k/2) + \dot{\theta}^2 \sin\theta(2\cos\theta - 1)$$

$$= \frac{3g(1 + k)\sin\theta}{a + b}.$$

Obtain $\dot{\theta}$ in terms of θ. [*Note:* The moment of inertia of a uniform cylinder about its axis is $\frac{1}{2}$(mass) $\times$ (radius)^2.]

8th British Mathematical Olympiad, 1972

1. S is a set of elements $\{a, b, \ldots\}$ each pair of which are distinct. R is a relation holding or not holding between every ordered pair of distinct elements of S according to the following rules:
   (a) either aRb or bRa, but not both;
   (b) if aRb and bRc, then cRa.
   Find the largest number of elements that S can have, proving your result.

2. Let a, b, c, and d be integers, with $a \neq b$. Show that there are at most four different points on the hyperbola

$$(x + ay + c)(x + by + d) = 2$$

each of the co-ordinates of which are integers. Find a set of conditions which are both necessary and sufficient for there to be four such points.

3. In a plane, two circles of unequal radii intersect at A and B, and through an arbitrary point P a straight line L is to be constructed so that the two

circles intersect equal chords on L. By considering distances from the point of intersection of L with AB (produced if necessary), or otherwise:
(a) find a 'ruler and compass' construction for L; and
(b) state the region of the plane in which P must lie for the construction to be possible.

4. A point P is on a smooth curve in a plane containing two fixed points A and B; $PA = a$, $PB = b$, and $AB = c$. The angle $APB = \theta$, where $c^2 = a^2 + b^2 - 2ab \cos \theta$. Show why

$$\sin^2\theta(\mathrm{d}s)^2 = (\mathrm{d}a)^2 + (\mathrm{d}b)^2 - 2(\mathrm{d}a)(\mathrm{d}b)\cos\theta,$$

where s measures the distance along the curve. The point P moves along the curve so that at time t during the interval $(\frac{1}{2}T < t < T)$ its distance from A and B are given, respectively, by $h \cos(t/T)$ and $k \sin(t/T)$, where h, k, and T are positive constants. Prove that the speed of P during this interval varies as cosec θ.

5. In a right circular cone the semi-vertical angle of which is θ, a cube is placed so that four of its vertices are on the base and four on the curved surface. Prove that as θ varies the maximum value of the ratio of the volume of the cube to the volume of the cone occurs when $\sin \theta = \frac{1}{3}$.

6. $T_k = k - 1$, for $k = 1, 2, 3, 4$. For all $k \geqslant 3$, $T_{2k-1} = T_{2k-2} + 2^{k-2}$ and $T_{2k} = T_{2k-5} + 2^k$. Show that, for all $k > 0$,

$$1 + T_{2k-1} = \left[\tfrac{12}{7}2^{k-1}\right] \qquad \text{and} \qquad 1 + T_{2k} = \left[\tfrac{17}{7}2^{k-1}\right],$$

where $[x]$ denotes the greatest integer not exceeding x.

7. Starting from $p_1 = 2$, $q_1 = 1$, $r_1 = 5$, and $s_1 = 3$, sequences of integers are formed by using the following relations, which hold for all positive integral n:

$$p_{n+1} = p_n^2 + 3q_n^2, \qquad q_{n+1} = 2p_n q_n, \qquad r_n = p_n + 3q_n, \qquad s_n = p_n + q_n.$$

Prove that:
(a) $p_n/q_n > \sqrt{3} > r_n/s_n$;
(b) p_n/q_n differs from $\sqrt{3}$ by less than $s_n/(2r_n q_n^2)$.

8. Three children, A, B, and C, build cairns of stones on a beach and try to knock down each other's cairns. Each time a child aims at a cairn, he has a fixed chance of hitting it: A has a chance of $\frac{3}{4}$ of hitting the cairn at which he aims, B a chance of $\frac{2}{3}$, and C a chance of $\frac{1}{2}$. Shots aimed at one cairn do not hit another. A player whose cairn is hit is out of the game. A game is won by being the only remaining player. The rule of play is for the players in the game to throw simultaneously, choosing with equal probability between his two opponents' cairns (when there are two). Find the probability that C wins, with A being eliminated in the first round.

9. A rocket, free of gravitational forces, is accelerating in a straight line. Its mass is M and that of its fuel is $m = m_0 e^{-kt}$. The relative velocity of escape of the fuel is $v = v_0 e^{-kt}$. If m_0 is small compared with M, show that the terminal velocity of the rocket is approximately $\frac{1}{2}m_0 v_0/M$ greater than its initial velocity.

7th British Mathematical Olympiad, 1971

1. (a) Factorize $(x+y)^7 - (x^7 + y^7)$.
   (b) Prove that there is no integer n for which $2n^3 + 2n^2 + 2n + 1$ is a multiple of 3.

2. $x_1 = 9$ and $x_{r+1} = 9^{x_r}$ for all positive integral r. Prove that the last two digits of x_3 written to base 10 are the same as the last two digits of x_4 written to base 10. What are these two digits?

3. Of a regular polygon of $2n$ sides there are n diagonals which pass through the centre of the inscribed circle. The angles which these diagonals subtend at two given points A and B on the circumference are $a_1, a_2, a_3, \ldots, a_n$ and $b_1, b_2, b_3, \ldots, b_n$. Prove that

$$\sum_{i=1}^{n} \tan^2 a_i = \sum_{i=1}^{n} \tan^2 b_i.$$

4. Given a set of $n+1$ positive integers, none of which exceeds $2n$, prove by induction or otherwise that at least one member of the set must divide another member of the set.

5. The triangle ABC has angles A, B, and C, in descending order of magnitude. Circles are drawn such that each circle cuts each side of the triangle internally in two real distinct points. The lower limit to the radii of such circles is the radius of the incircle of the triangle ABC. Show that the upper limit is not R, the radius of the circumcircle, and find this upper limit in terms of R, A, and B.

6. (a) $I(x) = \int_c^x f(x, u) du$, where c is a constant. Show why

$$I'(x) = f(x, x) + \int_c^x \frac{\partial f}{\partial x} du.$$

   (b) Find $\lim_{\theta \to 0} \cot \theta \sin(t \sin \theta)$.
   (c) $G(t) = \int_0^t \cot \theta \sin(t \sin \theta) d\theta$. Prove that $G'(\pi/2) = 2/\pi$.

7. Two real numbers h and k are given, such that $h > k > 0$. Find the probability that two points chosen at random on a straight line of length h should be at a distance of less than k apart.

8. A is a 3×2 matrix and B is a 2×3 matrix. The elements of each matrix are numbers. $AB = M$ and $BA = N$. det $M = 0$ and det $N \neq 0$. Also, $M^2 = kM$, where k is a number. Determine det N in terms of k, proving your result. You may assume the usual rules for combining matrices. [*Note:* k is not a matrix; kC is defined as the matrix each element of which is k times the corresponding element of C.]

9. Two uniform solid spheres of equal radii are so placed that one is directly above the other. The bottom sphere is fixed and the top sphere, initially at rest, rolls off. If the coefficient of friction between the two spherical surfaces is μ, show that slipping occurs when $2 \sin \theta = \mu(17 \cos \theta - 10)$, where θ is the angle that the line of centres makes with the vertical. [*Note:* The moment of inertia of a solid sphere of radius r and mass M about a diameter is $\frac{2}{5} Mr^2$.]

6th British Mathematical Olympiad, 1970

1. (a) Express the finite series

$$\frac{1}{\log_2 a} + \frac{1}{\log_3 a} + \ldots + \frac{1}{\log_n a}$$

   as a quotient of logarithms to base 2. Does the series converge as $n \to \infty$?

   (b) Evaluate the sum of the coefficients of the polynomial

$$p(x) = (1 + x - x^2)^3 (1 - 3x + x^2)^2,$$

   and the sum of the coefficients of its derivative.

2. Sketch the curve the equation of which is given by $x^2 + 3xy + 2y^2 + 6x + 12y + 4 = 0$. About which point does it possess some symmetry property?

3. You are probably aware of the remarkable fact that the three points of intersection of three appropriate pairs of trisectors of the internal angles of any triangle T form the vertices of an equilateral triangle t. Given a real constant $A > 0$, consider the class of all triangles T the area of which equals A. Find, with this class, the maximum value of the area of the triangle t. Discuss whether there is a minimum value.

4. S is any set of n positive integers the sum of which is not divisible by n. Prove that it is always possible to choose a subset of S, the sum of the members of which is divisible by n.

5. A cube of wood is to be sawn up into at least 300 separate pieces. What is the minimum number of sawcuts which need to be made? (Each sawcut is

planar, and it is assumed that pieces are not moved relative to each other
—so that the block remains intact as a cube—until after the last cut has
been made.)

6. In the region $|x| \leqslant a$, $y(x)$ is defined by the differential equation
$dy/dx = f(x)$, where $f(x)$ is a given even, continuous function of x.
Prove that:
   (a) $y(-a) - 2y(0) + y(a) = 0$;
   (b) $\int_{-a}^{a} y(x)dx = 2ay(0)$.
   A numerical step-by-step solution is calculated starting from the point
   $(-a, 0)$ using $2N$ equal x-wise steps, according to the usual scheme
   $g(x_{n+1}) = g(x_n) + \delta x \cdot g'(x_n)$. Prove that in general this solution will not
   satisfy the result (a) above. Suggest a modified step-by-step procedure
   which would ensure that (a) is satisfied.

7. The two base angles B and C of an isosceles triangle ABC are equal to
   $50°$. The point D lies on BC, so that $\angle BAD = 50°$, and the point E lies
   on AC, so that $\angle ABE = 30°$. Find $\angle BED$.

8. Eight electric light bulbs are arranged in a row, each controlled by its
   own on/off switch. State how many different patterns (or 'states') of lit
   bulbs there are (including the pattern of all bulbs off). Denote the N
   distinct states by S_i, $i = 1, 2, 3, \ldots, N$. Let n_i be the number of switches
   which have to be altered to change from S_i to S_{i+1} (with the convention
   that $S_{N+1} = S_1$). Find the smallest possible value of $n = \sum_{i=1}^{N} n_i$; that is,
   find the smallest value for the total number of switch-alterations which
   have to be made so as to run through all the possible states of the lights
   in some order, before returning to the initial state.

9. Find rational numbers x and y such that

$$\sqrt{(2\sqrt{3} - 3)} = x^{1/4} - y^{1/4}.$$

10. Unlimited supplies of straight rods (of negligible thickness) are available,
    the length of each rod being an integral multiple of 1 m. The rods are laid
    out on a large horizontal surface to form as many separate right-angled
    triangles as possible. Let $N(i)$ denote the number of different right-
    angled triangles the shortest side of which is a rod of length i metres.
    (For values of i for which no such right-angled triangle exists, we define
    $N(i) = 0$.)
    (a) Find some kind of 'formula' for $N(i)$.
    (b) Prove that, for any given positive integer M, there exists a value of i
        such that $N(i) > M$.
    (c) Discuss whether $N(i) \to \infty$ as $i \to \infty$.

5th British Mathematical Olympiad, 1969

1. Find the condition on the distinct real numbers a, b, and c such that $(x - a)(x - b)/(x - c)$ takes all real values for real values of x. Sketch two graphs of $(x - a)(x - b)/(x - c)$ to illustrate:
   (a) a case in which the condition is satisfied; and
   (b) a case in which the condition is not satisfied.

2. Find all the real solutions of $\cos x + \cos^5 x + \cos 7x = 3$.

3. A square piece of plywood is cut up into n^2 equal square pieces. These are then rearranged to make four rectangles with one square piece left over, in such a way that the nine dimensions of the five shapes (including the edge length of the square piece) are all different. Find the minimum value of n for this situation to be possible, and specify the shapes of the four rectangles for this value of n. State all other values of n for which the situation is possible, justifying your answer.

4. Find all possible pairs (a, b) of integers satisfying $a^2 - 3ab - a + b = 0$.

5. A long corridor of unit width has a right-angled corner in it. A rigid length of pipe (the thickness of which may be neglected) lies on, and is everywhere in contact with, the plane floor of the corridor. The length of the pipe (which may be curved) is defined as the straight-line distance between its two ends. Find the maximum length of pipe subject to the condition that it can be moved along both arms of the corridor and round the corner without leaving contact with the floor.

6. Prove that, if a, b, c, d, and e are positive integers, any common factor of $(ae + b)$ and $(ce + d)$ is also a factor of $(ad - bc)$.

7. Let $f(x)$ denote a real-valued function of the real variable x, not identically zero, and differentiable at $x = 0$.
   (a) If $f(x) \cdot f(y) = f(x + y)$ for all real values of x and y, prove that $f(x)$ is differentiable any number of times for all x, and that $\sum_{i=0}^{\infty} f(i) = 1/(1 - f(1))$ when $f(1) < 1$.
   (b) If $f(x) \cdot f(y) + f(x - y)$ for all real values of x and y, find $f(x)$.

8. A square has sides of length x. All four vertices of the square lie on the sides of a circumscribing triangle. The incircle of the triangle has radius r. Prove that $2r > x > \sqrt{2} \cdot r$.

9. Let $P(n)$ be the proposition that 'a polygon of n^2 sides can be drawn so that its vertices lie at the n^2 points of a square $n \times n$ array'. State, with proof, the truth or otherwise of $P(4)$ and $P(5)$. Formulate, with some justification, a conjecture about the truth of $P(n)$ for $n \geqslant 6$. [*Note:* A polygon satisfies the following common-sense properties: (a) no two sides

coincide; (b) no two sides intersect unless they are adjacent, non-parallel sides, in which case their intersection is one of the polygon's vertices.]

10. Describe a ruler-and-compass method of constructing an equilateral triangle the area of which equals that of a given triangle.

4th British Mathematical Olympiad, 1968

1. A circle C of unit radius rolls without slipping along the outside of the circle with centre the origin and radius 2 in the (x, y)-plane. A fixed point P of C is originally at the position $(2, 0)$. Find equations, giving the co-ordinates (x, y) of P in terms of a suitable parameter θ, as C rotates. Sketch the locus S described by P, showing all tangents parallel to the x- and y-axes.

2. Cows are put out to pasture when the grass has reached a certain height. Thereafter, as the cows eat the grass, the grass continues to grow as the pasture is consumed. If 15 cows can consume the grass in three acres of pasture in four days, while 32 cows can consume the grass in four acres in two days, how many cows will be required to consume the grass in six acres in three days?

3. The 'distance' d between two points (x_1, y_1) and (x_2, y_2) in the (x, y)-plane is defined by

$$d = |x_2 - x_1| + |y_2 - y_1|.$$

Using this notion of distance, find the locus of all points (x, y) satisfying $x \geqslant 0$, $y \geqslant 0$, and which are equidistant from the origin and from a fixed point (a, b) in the plane, where $a > b$. Distinguish the cases according to the signs of a and b.

4. Two spheres of radii a and b are tangent to each other and a plane is tangent to these spheres at different points. Find the radius of the largest sphere which can pass between the first two spheres and the plane.

5. Given x, y, and z such that

$$\sin x + \sin y + \sin z = 0$$

and

$$\cos x + \cos y + \cos z = 0,$$

prove that

$$\sin 2x + \sin 2y + \sin 2z = 0$$

and

$$\cos 2x + \cos 2y + \cos 2z = 0.$$

6. If $a_1, a_2, \ldots, a_7$ are integers, and $b_1, b_2, \ldots, b_7$ are the same integers rearranged, show that the value of

$$(a_1 - b_1)(a_2 - b_2) \ldots (a_7 - b_7)$$

is necessarily even.

7. A knock-out ping-pong tournament is arranged among n people and a new ball is used for each game. How many balls are needed?

8. A chord of length l divides the interior of a circle of radius r into two regions, D_1 and D_2. A circle S of maximal radius is inscribed in D_1; the area of the part of D_1 outside S is A. Show that A is greatest when the area of D_1 exceeds that of D_2, and when

$$l = \frac{16\pi r}{16 + \pi^2}.$$

9. Find the lengths of the sides of a triangle if its altitudes have lengths 3, 4, and 6.

10. The faces of a tetrahedron are formed by four congruent triangles. If α is the angle between a pair of opposite edges of the tetrahedron, show that

$$\cos \alpha = \frac{\sin(B - C)}{\sin(B + C)},$$

where B and C are the angles adjacent to one of these edges in a face of the tetrahedron.

11. The sum of the reciprocals of a set of n different positive integers is equal to one. If $n = 3$ show that there is only one such set and find it. Find such a set for $n = 4$, 5, and more generally for any value of $n > 3$.

12. Find, with proof, the maximum number of points which can be placed on the surface of a sphere of unit radius such that the distance between any two of the points is:
(a) at least $\sqrt{2}$;
(b) greater than $\sqrt{2}$.

3rd British Mathematical Olympiad, 1967

1. If α and β are the roots of $x^2 + px + 1 = 0$, and γ and δ are the roots of $x^2 + qx + 1 = 0$, show that

$$(\alpha - \gamma)(\beta - \gamma)(\alpha + \delta)(\beta + \delta) = q^2 - p^2.$$

2. By putting $y = tx$, or otherwise, draw the graph of $x^8 + xy + y^8 = 0$. Show clearly all turning values of x and y, and the behaviour of the graph when x and y become large.

3. (a) The lengths a and b of two sides of a triangle satisfy $a > b$. The lengths of the corresponding altitudes are h_a and h_b. Prove that $a + h_a \geqslant b + h_b$. Discuss all cases in which equality occurs.

   (b) If the circumcentre and the incentre of a triangle Δ coincide, prove that Δ is equilateral.

4. When is it possible to find points equidistant both from two given points P and Q and from a given straight line AB? When it is possible, show how to construct such points with ruler and compasses only.

5. Show that $f(t) = (t - \sin t)(\pi - t - \sin t)$ is increasing in the interval $0 < t < \pi/2$.

6. Find all real numbers x in the interval $0 \leqslant x \leqslant 2\pi$ such that

$$2 \cos x \leqslant |\sqrt{(1 + \sin 2x)} - \sqrt{(1 - \sin 2x)}| \leqslant \sqrt{2}.$$

   [*Note:* If t is any positive number, $\sqrt{t}$ is the positive square root of t. If x is any real number, $|x| = \sqrt{(x^2)}$.]

7. Four real numbers x_1, x_2, x_3, and x_4 are such that the sum of any one of them with the product of the other three is equal to 2. Find all possible solutions.

8. Find all non-negative integers n such that $5^n - 4^n$ is divisible by 61.

9. If α, β, and γ are the angles of a triangle with $\alpha\beta\gamma \neq 0$, show that $\cos^2\alpha + \cos^2\beta + \cos^2\gamma \geqslant \frac{3}{4}$. Find all such triangles for which equality holds. Show also that $\cos^2\alpha + \cos^2\beta + \cos^2\gamma$ does not have a greatest value.

10. Alan, Brian, Colin, and David collect stamps. Alan collects British stamps issued before 1900 and foreign stamps. Brian collects British stamps issued after 1900 and foreign special issues. Colin collects foreign stamps issued before 1900 and British special issues. David collects foreign stamps issued after 1900 and British special issues. Are there any stamps which no-one collects? Are there any stamps which everyone collects? What stamps could David give away that would be of interest to Alan but not to Brian?

11. In a certain town, the streets are arranged in a rectangular grid. A man wishes to go from one place to another place m streets east and n streets north. How many shortest paths are there?

2nd British Mathematical Olympiad, 1966

1. Find the least and the greatest values of $(x^4 + x^2 + 5)/(x^2 + 1)^2$ for real values of x.

2. Give the conditions under which all the roots of the equations

$$\pm\sqrt{(x-a)} \pm \sqrt{(x-b)} \pm \sqrt{(x-c)} = 0$$

are real, where a, b, and c are distinct real numbers.

3. Sketch the graph of $y^2 = x^2(x+1)/(x-1)$. Find all turning values of y, and describe the behaviour of the graph when x and y become large.

4. The points A, B, C, and D are four consecutive vertices of a regular polygon. If

$$\frac{1}{AB} = \frac{1}{AC} + \frac{1}{AD},$$

how many sides must the polygon have?

5. A square nut of side a is to be turned by means of a spanner, the hole in which consists of a regular hexagon of side b. Find the conditions on a and b for this to be possible.

6. Find the largest interval of values x for which the expression

$$y = \sqrt{(x-1)} + \sqrt{[x+24 - 10\sqrt{(x-1)}]}$$

has a constant value. [*Note:* $\sqrt{t}$ is defined for $t \geqslant 0$ to be that non-negative number whose square is t.]

7. Prove that $\sqrt{2}$, $\sqrt{3}$, and $\sqrt{5}$ cannot be terms of the same arithmetic progression.

8. The faces of a cube are coloured by six colours, so that each face has a different colour. Find how many cubes of distinct appearance can be produced in this way. Show also that 1680 regular octahedra of distinct appearance can be produced by colouring the eight faces of a regular octahedron with eight given colours so that each face has a different colour.

9. If α, β, and γ are the angles of a triangle, find:
   (a) the smallest possible value of $\tan(\alpha/2) + \tan(\beta/2) + \tan(\gamma/2)$;
   (b) the largest possible value of $\tan(\alpha/2)\tan(\beta/2)\tan(\gamma/2)$.

10. One hundred students, all of different heights, are arranged in a square of 10 rows by 10 columns. In each row the tallest student is selected, and the shortest of these tall students is labelled A. In each column the shortest student is selected, and the tallest of these short students is labelled B. If A and B are different persons, find which of them is the taller and why.

11. (a) Prove that, among any 52 integers, there must always exist two integers such that either the sum or the difference of these two integers is divisible by 100.

(b) Prove that, given any 100 integers, none of which is divisible by 100, it is possible to find two or more of these integers the sum of which is divisible by 100.

1st British Mathematical Olympiad, 1965

1. Sketch the graph of $y = (x^2 + 1)/(x + 1)$. Find all turning values, and describe the behaviour of the graph when x or y become large.

2. A pupil is swimming at the centre of a circular pond. At the edge of the pond there is a teacher, who wishes to catch the pupil, but who cannot swim. The teacher can run four times as fast as the pupil can swim, but not as fast as the pupil can run. Can the pupil escape from the teacher? Justify your answer.

3. Prove that, for every integer n:
   (a) $n^3 - n$ is divisible by 3;
   (b) $n^7 - n$ is divisible by 7;
   (c) $n^{13} - n$ is divisible by 13.

4. How many zeros are there at the end of the number 100!? [*Note:* $n!$ denotes the product of all the integers from 1 to n inclusive.]

5. Prove that the product of four consecutive integers is always one less than a perfect square.

6. Let $(2 + \sqrt{2})^n = P + F$, where P is an integer and $0 < F < 1$. Show that the positive integer n can be chosen so that $F > 0.999$.

7. Find the remainders upon dividing the polynomial $x + x^3 + x^9 + x^{27} + x^{81} + x^{243}$:
   (a) by $x - 1$;
   (b) by $x^2 - 1$.

8. Determine all values of the coefficient a for which the equations $x^2 + ax + 1 = 0$ and $x^2 + x + a = 0$ have at least one common root.

9. If a, b, and c are any three positive real numbers, prove that

$$(a + b)(b + c)(c + a) \geqslant 8abc.$$

Show that equality holds only when $a = b = c$.

10. A chord of length $\sqrt{3}$ divides a circle of unit radius into two regions. Find the rectangle of maximum area which can be inscribed in the smaller of these two regions.

Part III
Hints and outline solutions

Hints and outline solutions

It has come to me in a flash! One's intelligence may march about and about a problem, but the solution does not come gradually into view. One moment it is not. The next it is there.

William Golding, *Rites of Passage*

32nd British Mathematical Olympiad, 1996

> 1. Find as efficiently as possible all pairs (m, n) of positive integers satisfying the following two conditions:
> (a) two of the digits of m are the same as the corresponding digits of n, while the other two digits of m are both 1 less than the corresponding digits of n;
> (b) both m and n are four-digit squares.

(1) It may help to start by tackling the corresponding 'three-digit' problem.

> Determine all pairs (m, n) of positive integers satisfying:
> (a) two of the digits of m are the same as the corresponding digits of n, while the other digit of m is 1 less than the corresponding digit of n (as in, say, 263 and 273);
> (b) both m and n are three-digit squares.

If m has three digits then m must look like 'abc'. If n has two digits the same as m (and in the same positions as m), with the other digit one larger than for m, then n must look like

$$\text{'}ab(c + 1)\text{'}, \quad \text{or} \quad \text{'}\underline{\qquad}\text{'}, \quad \text{or} \quad \text{'}\underline{\qquad}\text{'}.$$

(2) If $m = \text{'}abc\text{'}$ and $n = \text{'}ab(c + 1)\text{'}$, then the two squares '$abc$' and '$ab(c + 1)$' differ by _____. Hence the only possibilities for m and n are '_____' and '_____', so m is not 'positive'; (anyway, '000' and '001' are not really 'three-digit' numbers). In the second case the two squares 'abc' and '$a(b + 1)c$' differ by exactly _____; a short check should convince you that there are no

solutions. In the third case the two squares differ by exactly _____ , and there is just one solution: namely $m =$ _____ , $n =$ _____ .

(3) Before moving on to the 'four-digit' problem, you should try to improve on the above trial-and-error approach. How can you be *sure* that there is no solution in the second case (that is, with $n - m = 10$)? How do you *know* that there is only one solution in the third case? The key to a more mathematical approach lies in writing the squares m and n as squares: $m = x^2$ and $n = y^2$. Then

$$n - m = y^2 - x^2 = (\underline{\hspace{2cm}})(\underline{\hspace{2cm}}).$$

There is more to this factorization than meets the eye, for the two factors on the RHS differ by $2x$, which is e*e*. Hence either:

(3.1) both factors must be o**, so $n - m$ is odd; or

(3.2) both factors must be e*e*, so $n - m$ is a multiple of *ou*.

Now check the three cases in (2) properly. In the first case,

$$n - m = 1; \quad \therefore \; y - x = 1 = y + x \;\Rightarrow\; y = \underline{\hspace{1cm}}, \; x = \underline{\hspace{1cm}}.$$

In the second case,

$n - m = 10$, which is e*e*, but not a multiple of *ou*;

hence there are no solutions.

In the third case,

$n - m = 100$;

$\therefore$ either $y - x = y + x =$ _____ , so $x = 0$ and m is not 'positive';
or $y - x =$ _____ , $y + x =$ _____ (since $y - x$ and $y + x$ must both be e*e*),
$\therefore$ $y =$ _____ , $x =$ _____ , so $m =$ _____ , $n =$ _____ is the only solution.

(4) If m and n both have four digits, and satisfy condition (a) in the question, then there are exactly *i* possible positions for the pair of digits common to both m and n. Hence, if m has the form '*abcd*', n has the form

'$ab(c + 1)(d + 1)$',   or   '_____',   or   '_____',   or   '_____',

or   '_____',   or   '_____'.

You must now analyse each of these six cases separately using the factorization in (3) above, and the conditions (3.1) and (3.2). (*Note:* In each case there may be no solution, or one solution, or more than one solution.)

2. A function f is defined for all positive integers and satisfies

$$f(1) = 1996,$$

and

$$f(1) + f(2) + \ldots + f(n) = n^2 f(n) \qquad \text{for all } n > 1.$$

Calculate the exact value of $f(1996)$.

(1) If you have never seen anything like this before, it seems reasonable to start calculating: $f(1) = 1996$ (given); $f(2) = \underline{\quad}$ (from $f(1) + f(2) = 2^2 f(2)$); $f(3) = \underline{\quad}$, and so on. Use this approach to find $f(2)$, $f(3)$, and $f(4)$.

(2) If you persevere, and keep your wits about you, you might just notice something interesting (though you must be careful not to let the numbers 1996, 1, 2, 3, 4, etc. obscure what is going on). If you are lucky, you may even be able to guess a value for $f(1996)$. But this would not answer the question! In mathematics, the word 'determine' means more than just 'guess' or 'find'; it means that you have to show exactly why your value is correct. In other words, you have to 'find the correct value, *and prove it is correct*'. For this, it is not the *values* of $f(2)$, $f(3)$, and so on, that matter, but their *form*. Thus it is important to express $f(2)$ and $f(3)$ in a form that reveals what is really going on:

$$f(2) = \frac{1}{(2^2 - 1)} \cdot f(1),$$

$$f(3) = \frac{1}{3^2 - 1} \cdot [f(1) + f(2)] = \frac{1}{3^2 - 1} \cdot \left[f(1) + \frac{1}{(2^2 - 1)} \cdot f(1) \right]$$

$$= \frac{1}{(2^2 - 1)} \cdot \frac{2^2}{(3^2 - 1)} \cdot f(1).$$

Write out the calculation which shows that

$$f(4) = \frac{1}{(2^2 - 1)} \cdot \frac{2^2}{(3^2 - 1)} \cdot \frac{3^2}{(4^2 - 1)} \cdot f(1).$$

(3) Now guess what you expect to be the corresponding expression for $f(n)$ in terms of $f(1)$, and *prove* that your guess is correct (by induction on n).
(4) Even at this stage it is important to resist the temptation simply to substitute $n = 1996$. Factorize each of the factors $(r^2 - 1)$ in the denominator of your (proven!) expression for $f(n)$, and cancel to obtain a greatly simplified formula for $f(n)$ in terms of n and $f(1)$. Finally, substitute $n = 1996$.

> 3. Let ABC be an acute-angled triangle, and let O be its circumcentre. The circle through C, O, and B is called S. The lines AB and AC meet the circle S again at P and Q respectively. Prove that the lines AO and PQ are perpendicular.

(1) The first thing to do in any geometry problem is to get out your ruler and compasses (and a sharp pencil), and draw a good diagram. As you construct the diagram, try to understand what it is that you have to prove. In your diagram, it is highly likely that the line segments AO and PQ do not even meet! So perhaps you should extend the line segment AO to meet PQ at the point X (say). You have to show that $\angle AXP$ is a right angle. You would scarcely be asked to 'prove' this if it was obvious, so you must expect to have to do something for yourself.

(2) The simplest thing you could hope for is that the required result '$\angle AXP = 90°$' may be equivalent to something which is a little more obvious. In the question there is no mention of the point X; all of the information given is about the points A, B, C, O, P, and Q. Use the triangle AXP to reformulate the required result '$\angle AXP = 90°$' into an equivalent statement involving only $\angle BPQ$ and $\angle BAO$.

(3) Somewhere you must expect to use the fact that the four points B, P, Q, and C lie on a *i***e, and so form a ****i* quadrilateral. Use this to reformulate the required result into a statement involving $\angle BCA$ and $\angle BAO$ only.

(4) One reason why (3) is a good move is that you now know that the required result '$\angle AXP = 90°$' is equivalent to a statement about angles in the *original triangle ABC*. Read the question again carefully. What does the fact that 'O is the circumcentre of triangle ABC' tell you about triangle OAB? What does it tell you about triangle OBC? What does it tell you about triangle OCA? (If you did not mark in the line segments BO and CO when you drew the original diagram, it should now be clear that this was a mistake. You should always expect to have to mark *important* lines or points which are not explicitly mentioned in the question—although you must be selective, since if you mark too many points and lines the diagram becomes a mess.) How does this prove what you want to prove about $\angle BAO + \angle BCA$?

[Alternatively, $\angle BAO = \angle XAB = \angle ABO$ (since $AO = BO$, so $\triangle AOB$ is isosceles); $\angle AOB = 2 \times \angle ACB$ (since O is the centre of the circle through A, B, and C). Hence $180° = \angle XAB + \angle ABO + \angle AOB = 2(\angle BAO + \angle BCA)$.]

> 4. For any real number x, let $[x]$ denote the greatest integer which is less than or equal to x. Define $q(n) = \left[\dfrac{n}{[\sqrt{n}]} \right]$ for $n = 1, 2, 3, \ldots$. Determine all positive integers n for which $q(n) > q(n+1)$.

(1) One of the amazing things about mathematics is the way one can start out in a complete fog, yet still manage to fumble one's way to the required goal! At first sight it is not at all clear what this question is about, nor how one is meant to begin. The unfamiliar '[x]' makes the whole thing *look* much worse. But don't just sit around feeling glum: start by calculating a few values $q(1), q(2), q(3), q(4), q(5), \ldots$, just to see how this strange looking function $q(n)$ behaves. When $n = 1$, $\sqrt{1} = \underline{\quad}$; so $[\sqrt{1}] = \underline{\quad}$; so $1/[\sqrt{1}] = \underline{\quad}$; so $q(1) = [1/[\sqrt{1}]] = \underline{\quad}$. Now do the same to find $q(2)$, $q(3)$, $q(4)$, etc.

(2) Before you can answer the question you need an *idea* about how to begin. The first thing that you are almost bound to notice is that the nasty looking denominator $[\sqrt{n}]$ *jumps* whenever n moves from a number just below a perfect **ua*e to the next **ua*e. So suppose that we choose a square m^2 and concentrate on values of n between two successive perfect squares m^2 and $(m + 1)^2$; that is, with $n \geqslant m^2$ but $n < (m + 1)^2$. Then $\sqrt{n} \geqslant \underline{\quad}$ and $\sqrt{n} < \underline{\quad}$. Hence $[\sqrt{n}] = \underline{\quad}$. For such values of n, the exact value of $q(n) = [n/m]$ will depend on where n is in the interval $m^2 \leqslant n < (m + 1)^2$:

- if $m^2 \leqslant n < m^2 + m$, then $q(n) = [n/m] = \underline{\quad}$;
- if $m^2 + m \leqslant n < m^2 + 2m$, then $q(n) = [n/m] = \underline{\quad}$;
- if $n = m^2 + 2m$, then $q(n) = [n/m] = \underline{\quad}$.

Thus for all of these values, $q(n)$ increases as n increases.

(3) Hence the only value of n for which we may find $q(n) > q(n + 1)$ is $n = \underline{\quad}$. Calculate $q(m^2 + 2m)$ and $q((m + 1)^2)$.

(4) This more or less finishes off the question. But it may be worth reflecting on the way in which a question that *seemed* so inaccessible a little while ago now seems so simple.

---

5. Let a, b, and c be positive real numbers. Prove that:
   (a) $4(a^3 + b^3) \geqslant (a + b)^3$;
   (b) $9(a^3 + b^3 + c^3) \geqslant (a + b + c)^3$.

If you have a clever idea about how to solve both parts at once, you will probably not need this outline solution. So I shall begin as though you have *not* yet solved the problem.

(1) The first inequality (a) almost invites you to multiply out the bracket on the RHS and collect up terms. Do this: you should be able to write out the expansion of $(a + b)^3$ without doing any work, using the coefficients 1, 3, 3, and 1 from Pascal's triangle. ('Multiplying out' is often the *wrong* thing to do; but unless you have a better idea, the cubed terms on both sides of (a), and the fact that all terms land up with the same coefficient '3', suggest that it might lead here to a simpler version of the problem.)

(2) This shows that the required inequality is equivalent to proving that

$$a^3 - a^2b - ab^2 + b^3 \geqslant 0 \qquad \text{whenever } a, b > 0.$$

The LHS of this inequality is just asking to be factorized (the first and third terms have a common factor a, while the second and fourth terms have a common factor b; both the resulting terms then have a common factor $a^2 - b^2$). If you complete this factorization, the LHS has three factors—one of which is positive, while the other two are equal; so we have '(something positive) times (a perfect square)', which must therefore be $\geqslant 0$. Hence the whole expression is always $\geqslant 0$, as required.

(3) The steps in (1) and (2) above start out with the inequality that is to be proved, and transform it into an equivalent inequality that you *can* prove. While it is possible to write this out correctly (provided that you are very careful to stress that 'a and b satisfy $4(a^3 + b^3) \geqslant (a + b)^3$ *if and only if* $a^3 - a^2b - ab^2 + b^3 \geqslant 0$'), it is better to make a habit of *never* starting out from what you have to prove. Instead, you should start from something that is indisputably true and move step by step to deduce what you wish to prove. For example:

'Suppose that $a, b > 0$. Then $a + b > 0$, and $(a - b)^2 \geqslant 0$.

$\therefore (a + b)(a - b)^2 = a^3 - a^2b - ab^2 + b^3 \geqslant 0$

$\therefore 3(a^3 - a^2b - ab^2 + b^3) \geqslant 0$

$\therefore 4(a^3 + b^3) \geqslant a^3 + 3a^2b + 3ab^2 + b^3 = (a + b)^3.$'

[An alternative approach to (a) is to notice that $a^3 + b^3$ can itself be factorized (see the 'Algebra' section in 'A little useful mathematics'). One (positive) factor cancels with one of the factors of $(a + b)^3$. It follows that the required inequality is equivalent to $4(a^2 - ab + b^2) \geqslant (a + b)^2$; which simplifies to $a^2 - 2ab + b^2 \geqslant 0$, or $(a - b)^2 \geqslant 0$.]

(4) Unfortunately—as you have probably discovered if you are reading this hint in search of inspiration—it is not easy to make these simple-minded methods work when the number of variables increases from two (a, b) to three (a, b, c)! But that only means that you need an *idea*, rather than blind calculation. One bright idea you might think of is to try to use part (a). You know that a, b, and c are all positive. Thus you can safely use the result of (a) for each of the three paris 'a, b', 'b, c', and 'c, a' separately:

$$\therefore 4(a^3 + b^3) \geqslant (a + b)^3, \qquad 4(b^3 + c^3) \geqslant (b + c)^3, \qquad 4(c^3 + a^3) \geqslant (c + a)^3.$$

Now add these three inequalities, simplify the LHS, and multiply out the RHS. The resulting inequality is certainly not quite what you want, but it may come in handy later on.

(5) You want to *prove* the inequality $9(a^3 + b^3 + c^3) \geqslant (a + b + c)^3$. To succeed, you must certainly not start by *assuming* what you want to prove; but it is a good idea to start with the LHS '$9(a^3 + b^3 + c^3)$', and to try to find a succession of true inequalities that will lead you to the RHS. So, start with $9(a^3 + b^3 + c^3)$ and use the inequality that you proved in (4) to show that

$$9(a^3 + b^3 + c^3) \geqslant 3(a^3 + b^3 + c^3 + a^2b + b^2a + b^2c + c^2b + c^2a + a^2c).$$

(6) Now work in rough to compare the RHS of the inequality proved in (5) with the RHS $(a + b + c)^3$ of the inequality that you are trying to prove. Show that

$$3(a^3 + b^3 + c^3 + a^2b + b^2a + b^2c + c^2b + c^2a + a^2c) \geqslant (a + b + c)^3$$

is true, *provided that* $2a^3 + 2b^3 + 2c^3 \geqslant 6abc$ holds. Then finish off the question.

31st British Mathematical Olympiad, 1995

> 1. Find all squares ending in three 4s. Show that no square ends in four 4s.

(1) The first number that ends in three 4s is 444, which is not a square (since it lies between $21^2 = 441$ and $22^2 = 484$). The very next candidate is _____, and this just happens to be equal to (_____)^2. Thus you should certainly have managed to find at least one such square.

(2) The challenge is to find *all* squares N^2 which end in three 4s. The first thing to realize is that you only need to worry about the *last three digits* of N (since if $N = 1000k + x$, then $N^2 = (1000k + x)^2 = 1000(\underline{\hspace{2cm}}) + x^2$, so the only bit which affects the last three digits of N^2 is the final term x^2). You can now use one of two approaches.

(3) For the *last* digit of x^2 to be a 4, the last digit of x must be __ or __. The last two digits of x^2 are determined by the square of the last two digits of x (Why?). If you square in turn 02, 12, 22, 32, 42, 52, 62, 72, 82, 92 and 08, 18, 28, 38, 48, 58, 68, 78, 88, 98 (no calculators!) you find that only __, __, __, and __ give two 4s. You can now repeat this exercise to discover which endings x have squares ending in *three* 4s. (This approach is rather messy. See if you can streamline it to make it more mathematical.)

(3$'$) Alternatively, you know from (1) that 38^2 ends in three 4s. Hence, if N^2 also ends in three 4s, then $N^2 \equiv 38^2 \pmod{1000}$. Therefore $N^2 - 38^2 = (N - 38)(N + 38) \equiv 0 \pmod{1000}$, or $1000 \,|\, (N - 38)(N + 38)$. Now HCF$(N - 38, N + 38) = $ HCF$(N - 38, (N + 38) - (N - 38)) = $ HCF$(N - 38, 76)$. Hence

HCF($N - 38, N + 38$) must divide $76 = 2^2 \times 19$. Since we know that $2^3 \times 5^3 = 1000 \mid (N + 38)(N - 38)$, it follows that $5^3 \mid N - 38$ or $5^3 \mid N + 38$; moreover, both $N + 38$ and $N - 38$ must be divisible by 4 (Why?). Hence

$$N^2 \text{ ends in three 4s} \Rightarrow \text{ either } N + 38 \text{ or } N - 38 \text{ is divisible by } 4 \times 125,$$

$$\Rightarrow N = 500k \pm 38 \text{ for some } k.$$

Conversely, if $N = 500k \pm 38$, then N^2 ends in three 4s (Why?).
(4) Finally, you should check that, when N has this form, then N^2 never ends in four 4s.

[A better method here is to observe the following. If N^2 ends in four 4s, then $N = 2M$ is even, and M^2 ends in at least two 1s (Why only *two*?). But then $M^2 \equiv 11 \pmod{100}$, so M^2 would be congruent to 3 (mod 4), which is impossible!]

---

2. *ABCDEFGH* is a cube of side 2.
   (a) Find the area of the quadrilateral *AMHN*, where *M* is the midpoint of *BC*, and *N* is the midpoint of *EF*.
   (b) Let *P* be the midpoint of *AB*, and *Q* the midpoint of *HE*. Let *AM* meet *CP* at *X*, and let *HN* meet *FQ* at *Y*. Find the length of *XY*.

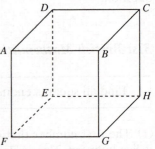

---

(1) Start with (a). We can only calculate 'area' for two-dimensional figures, so you should first show that the plane through *A*, *M*, and *H* passes through *N* too. (It is enough to show that *AMHN* is a parallelogram.) $\triangle ABM$ has a *i*** a***e at ____, so we can use Pythagoras to calculate $AM = $ ____. Similarly, we can find *MH*, *HN*, and *NA*. All four sides of *AMHN* are e*ua*, so *AMHN* is a _____. (Do not fall into the trap of assuming that *AMHN* is a **ua*e!). The diagonal *AH* has length ____ (Why?), and *MN* has length ____ (Why?); hence area($AMHN$) = ____.

[Alternatively, represent *AM* and *AN* as vectors and take the cross product.]
(2) Now for (b). Take *A* as the origin, and find the co-ordinates of *X* (using 2-D geometry in *ABCD*). Then find the co-ordinates of *Y*. Then use Pythagoras (in 3-D) to calculate the length of *XY*.

[Or use vectors: $\vec{AB} = 2\mathbf{i}$, $\vec{AF} = 2\mathbf{j}$, $\vec{AD} = 2\mathbf{k}$, and $\vec{AX} = \lambda(2\mathbf{i} + \mathbf{k}) = \mathbf{i} + \mu(2\mathbf{k} + \mathbf{i})$, so $\mu = $ ____, and $\vec{AX} = $ __ $\mathbf{i} + $ __ $\mathbf{k}$. Similarly, $\vec{AY} = $ __ $\mathbf{j} + $ (__ $\mathbf{k} + $ __ $\mathbf{i}$); $\therefore$ $\vec{XY} = $ __ $\mathbf{k} + $ __ $\mathbf{j} + $ __ $\mathbf{i}$, so the required length $XY = $ ____ .]

3. Maximize (a) $x^2y - y^2x$ and (b) $x^2y + y^2z + z^2x - x^2z - y^2x - z^2y$,
   when $0 \leqslant x, y, z \leqslant 1$.

(1) Start with (a). It is not hard to discover what *seems* to be the maximum value; namely, ____, when $x =$ ____ and $y =$ ____. The challenge is to *prove* that no other x and y can do better!

(2) Here is one way. Since $0 \leqslant x \leqslant 1$, and $xy \geqslant 0$, $x^2y = x \cdot xy \leqslant xy$. Thus $x^2y - y^2x = x \cdot xy - y^2x \leqslant xy - y^2x = x(y - y^2) \leqslant y - y^2$ (since $x \leqslant 1$ and $y - y^2 = y(1 - y) \geqslant 0$). Finally, observe that $y - y^2 = \frac{1}{4} - (\frac{1}{2} - y)^2 \leqslant \frac{1}{4}$ (since the squared term is always $\geqslant 0$). Moreover, $x^2y - y^2x = \frac{1}{4}$ when $x =$ ____, $y =$ ____ .

[Alternatively, for each *fixed* x ($0 \leqslant x \leqslant 1$), differentiating with respect to y shows that $x^2y - y^2x$ has maximum value ____ at $y =$ ____. Hence the maximum over all x is at $x =$ ____, $y =$ ____ .]

(3) Now for part (b). The expression can take positive values (for example, when $x =$ ____, $y =$ ____, $z =$ ____); so the maximum is certainly positive. When $x = y$ (or $y = z$, or $z = x$), the expression equals ____; hence the remainder theorem gives

$$(*) \quad x^2y + y^2z + z^2x - x^2z - y^2x - z^2y = (x - y)(y - z)(x - z)$$

$$= (x - y)(z - y)(z - x) = (y - z)(z - x)(y - x).$$

For a maximum, either:

(i) all brackets in $(x - y)(y - z)(x - z)$ are positive, so $x > y > z$; or

(ii) exactly two brackets in $(x - y)(y - z)(x - z)$ are negative, so $z \geqslant x \geqslant y$ or $y \geqslant z \geqslant x$ and all of the brackets in one of the other factorizations in $(*)$ are positive!

Relabelling the variables if necessary, you may assume that $x \geqslant y \geqslant z$. Then $(x - y)(y - z)(x - z) \leqslant (1 - y)(y - z)(1 - z) = a \cdot (b - a) \cdot b = b^2a - a^2b$, where $a = 1 - y \in [0, 1]$, and $b = 1 - z \in [0, 1]$. Now apply part (a) (with $x = b$, $y = a$) to obtain a maximum when $a =$ ____, $b =$ ____ .

[There are many other correct ways. But beware: most arguments are flawed!]

4. Triangle ABC has a right angle at C. The internal bisectors of angles BAC and ABC meet BC and CA at P and Q, respectively. The points M and N are the feet of the perpendiculars from P and Q to AB. Find angle MCN.

(1) Draw an accurate, decent-sized diagram.
(2) You should always expect to have to add important lines and points that

are not mentioned explicitly in the question. The segments PM and QN are both perpendicular to AB; this should suggest that it might be a good idea to draw the perpendicular CL from C to AB. (Then $\triangle ACB$, $\triangle ANQ$, $\triangle ALC$, $\triangle CLB$, and $\triangle PMB$ are *i*i*a* (Why?). Hence $AN:AL = AQ:AC$.)

(3) BQ bisects the angle ABC. What does this tell you about the triangles BQC and BQN? (Explain.) What can you conclude about triangle QNC? Use this (and the fact that QN and CL are parallel) to prove that CN bisects $\angle LCQ$.

(4) Prove that CM bisects $\angle LCP$.

(5) Deduce that $\angle MCN = \angle MCL + \angle LCN = \frac{1}{2}\angle LCB + \frac{1}{2}\angle LCA$
$$= \frac{1}{2}(\angle LCB + \angle LCA) = \underline{\qquad}.$$

> 5. The seven dwarfs walk to work each day in single file, with heights alternating *up–down–up–down–* ... or *down–up–down–up–* .... If they all have different heights, for how long can they continue like this with a new order every day? For how long could they continue if Snow White always came along too?

(1) We need a recurrence for $S(n)$—the number of arrangements with n dwarfs. The answer to the first and second parts will then be $S(7)$ and $S(8)$ respectively. Let the dwarfs be numbered $1, 2, 3, \ldots, n$, where dwarf i may be assumed to have height i. Let $U(n)$ be the number of possible *up–down–* ... arrangements, and $D(n)$ the number of *down–up–* ... arrangements for n dwarfs. Suppose that you are given some *up–down–up–* ... arrangement. If, for each i $(1 \le i \le n)$, you replace dwarf i by dwarf $(n+1) - i$, the *up–down–up–* ... arrangement becomes a *down–up–down–* ... arrangement —and conversely. Hence $U(n) = D(n)$. (This simple proof has to be given; the result isn't obvious.) When $n \ge 2$, $U(n) = D(n) = S(n)/2$; when $n = 0$ or 1 one has to be slightly careful, since then $U(n) = D(n) = S(n) = 1$.

(2) Count the number $S_i(n)$ of possible (*up–down–* ... or *down–up–* ...) arrangements in which the tallest dwarf 'n' is in the $(i + 1)$th position. To obtain such an arrangement, you must first choose i dwarfs from $\{1, 2, \ldots, n - 1\}$ to go in the first i positions; then order these i dwarfs (*up–down–* ... or *down–up–* ...); finally, you must order the remaining $n - (i + 1)$ dwarfs (*up–down–* ...). Since $U(i) = D(i)$, the formula is the same irrespective of whether the first i dwarfs go '*up–down–* ...' or '*down–up–* ...'; namely,

$$S_i(n) = \binom{n-1}{i} \times U(i) \times U(n - i - 1).$$

(3) When $i > 1$, $U(i) = S(i)/2$; but when $i = 0$ or 1, $U(i) = S(i) = 1$. Hence we may use

$$S(n) = \sum_{i=0}^{n-1} S_i(n),$$

with starting values $U(0) = S(0) = 1$ and $U(1) = S(1) = 1$, to calculate $S(2)$, $S(3)$, etc.:

$$S(2) = \binom{1}{0} U(0)U(1) + \binom{1}{1} U(1)U(0) = 1 \cdot 1 \cdot 1 + 1 \cdot 1 \cdot 1 = \underline{\quad};$$

$$S(3) = \binom{2}{0} U(0)U(2) + \binom{2}{1} U(1)U(1) + \binom{2}{2} U(2)U(0)$$
$$= 1 \cdot 1 \cdot 1 + 2 \cdot 1 \cdot 1 + 1 \cdot 1 \cdot 1 = \underline{\quad};$$

$$S(4) = \binom{3}{0} U(0)U(3) + \binom{3}{1} U(1)U(2) + \binom{3}{2} U(2)U(1)$$
$$+ \binom{3}{3} U(3)U(0) = \underline{\quad};$$

$$S(5) = \qquad\qquad ;$$
$$S(6) = \qquad\qquad ;$$
$$S(7) = \qquad\qquad ;$$
$$S(8) = \qquad\qquad .$$

30th British Mathematical Olympiad, 1994

> 1. Starting with any three-digit number n (such as $n = 625$), we obtain a new number $f(n)$ which is equal to the sum of the three digits of n, their three products in pairs, and the product of all three digits.
> (a) Find the value of $n/f(n)$ when $n = 625$. (The answer is an integer!)
> (b) Find all three-digit numbers n such that the ratio $n/f(n) = 1$.

(1) Part (a) is only included to help you to check that you have read the question. (If you calculate $n/f(n)$ and obtain an integer, then you have probably got the right idea. Otherwise, go back and read the question again more carefully.)

(2) Now for part (b). In the BMO 'find' means 'find them all, and prove that you have found them all'. By all means work in rough to find what you think is the complete list. But do not then think that this solves the problem: it is only a beginning. The first mathematical step is straightforward algebra: if $n = \text{'}abc\text{'} = 100a + 10b + c$, then $f(n) = abc + ab + bc + ca + a + b + c$. Hence $n = f(n)$ precisely when

$$99a + 9b = abc + ab + bc + ca.$$

That is,

$(*)$ $\qquad\qquad n = f(n) \implies (9 - c)b = a(bc + b + c - 99).$

Now $b, c \leqslant 9$, so $bc + b + c - 99 \leqslant$ ___ . Since $a \geqslant 0$; it follows that the RHS of $(*)$ is $\leqslant$ ___ . But $b \geqslant 0$, and $c \leqslant 9$ (so $9 - c \geqslant 0$); hence the LHS of $(*)$ is $\geqslant$ ___ . Thus $f(n) = n$ is true if and only if both sides of $(*)$ are *equal* to ___ . Since $a \neq 0$ (as n is a *three*-digit number), we must have $bc + b + c = 99$; hence $b = c = 9$. Thus $n = 199, 2$___ , ___ , ___ , ___ , ___ , ___ , ___ , or ___ .

[Alternatively, use $(*)$ to solve for c. Use this expression for c to show $c \geqslant 9$, with equality only if $b = 9$. Hence conclude that $b = 9 = c$.]

---

2. In triangle ABC the point X lies on BC.
   (a) Suppose that $\angle BAC = 90°$, that X is the midpoint of BC, and that $\angle BAX$ is one third of $\angle BAC$. What can you say (and prove!) about triangle ACX?
   (b) Suppose that $\angle BAC = 60°$, that X lies one third of the way from B to C, and that AX bisects $\angle BAC$. What can you say (and prove!) about triangle ACX?

---

(1) In your attempt at part (a) you may not have taken the injunction '(and prove!)' seriously enough. 'Prove' in mathematics means 'lay out logically, and explain each step clearly in terms of basic facts or theorems'. Do not be satisfied with baby language: for example, rather than 'X is halfway along BC so X must be halfway up (from AB to C) and halfway across (from B to AC)', you should use $*i*i*a* **ia***e*$. (If you join the midpoint X of BC to the midpoint M of AC, you must explain logically why XM is perpendicular to AC. And if you choose M on AC so that XM is perpendicular to AC, you must explain why M has to be the midpoint of AC.)

(2) One possible approach to part (a) is to use the sine rule to find AB from triangle AXB in terms of $\sin(\angle AXB)$, then find AC from triangle AXC in terms of $\angle AXC$. [Alternatively, use the area formula '$(\frac{1}{2})ab \sin C$' in each triangle.] Hence conclude that $AB : AC = \sqrt{3}$, so $\tan(\angle ACB) =$ ___ , whence $\angle ACX =$ ___ . Hence $\triangle ACX$ is isosceles with base angles equal to 60°, so it must in fact be $e*ui*a*e*a*$. (*Note:* It is not enough to derive a trig equation relating certain angles in the diagram, and then to simply 'assert' that your favourite solution is the only possible solution. If you use such an approach, you must *prove* that there is only one possible solution.)

[Alternatively, you could find AX from both triangles and conclude that $\tan(\angle ABX) = 1/\sqrt{3}$. Or you could observe that the circumcircle of $\triangle ABC$ has diameter BC (since $\angle BAC$ is a right angle), so X is the centre. Hence $AX = BX = CX$, and so on.]

(3) Part (b) is related to part (a). But this does not mean that you can assume what needs to be *proved* (namely that $\angle B = 90°$) and work backwards! You could use the area formula '$(\frac{1}{2})AB \cdot AX \cdot \sin$ ___ $: (\frac{1}{2})AC \cdot AX \cdot \sin$ ___ $= 1 : 2$' to deduce that $AC : AB =$ ___ . [The sine rule in triangles AXB and AXC

gives the same result.] The cosine rule on $\triangle ABC$ (for $\angle A$) then gives $BC:AB =$ ____ , and the converse of Pythagoras (or the cosine rule on $\triangle ABC$ (for $\angle B$)) gives $\angle ABC =$ ____ . It is then not hard to show that $AX = XC$, or that $\triangle AXC$ is _____ . (Your final solution must be laid out carefully, giving reasons for each step.)

---

3. The sequence of integers $u_0, u_1, u_2, u_3, \ldots$ satisfies $u_0 = 1$ and

$$u_{n+1} u_{n-1} = ku_n \qquad \text{for each } n \geqslant 1,$$

where k is some fixed positive integer. If $u_{2000} = 2000$, determine all possible values of k.

---

(1) Think carefully before writing down lots of equations (or you may just go round in circles). Make sure you understand, and avoid, the following common mistakes.

(a) Once *one* term repeats, you *cannot* conclude that the whole sequence must repeat. Instead you have to show not only that $u_6 = u_0$, but also that $u_7 = u_1$ (since it is u_6 and u_7 together that determine u_8, and so on).

(b) There are several pieces of information in the question for you to juggle. You must not only take them all into account, but you must *show the reader how you are doing this*. It is not enough simply to declare 'The only possible values of k are _____ '.

(2) Let $u_1 = x$. Then $u_2 =$ ____ , $u_3 =$ ____ , $u_4 =$ ____ , $u_5 =$ ____ , $u_6 =$ ____ , and $u_7 =$ ____ .

(3) Since each term is determined by the two preceding terms, and $u_6 = 1 = u_0$ and $u_7 = u_1 = x$, the sequence must recur. Hence $u_0 = u_6 = u_{12} = \ldots$, so $u_{2000} = u_?$.

(4) But $u_2 = kx$, where k and x are integers, and k is positive. Hence $u_{2000} = 2000$ implies that k and x must both divide 2000. Now $2000 = 2^4 \times 5^3$, so k and x are both of the form $2^a \cdot 5^b$. Moreover, $u_5 = k/x$ must be an integer, so $2000 = 2^4 \cdot 5^3 = (k/x) \cdot x^2$. Therefore $x =$ ____ , ____ , ____ , ____ , ____ , or ____ , and $k =$ ____ , ____ , ____ , ____ , ____ , or ____ . Each of these pairs x, k gives rise to a sequence of integers satisfying the conditions in the question.

---

4. The points Q and R lie on the circle γ, and P is a point such that PQ and PR are tangents to γ. The point A lies on the extension of PQ, and γ' is the circumcircle of triangle PAR. The circle γ' cuts γ again at B, and AR cuts γ at the point C. Prove that $\angle PAR = \angle ABC$.

(1) This is in fact the easiest question on the paper; but that does not automatically make it popular! Very little geometry is taught nowadays; so if you are interested in mathematics, here is a truly wonderful branch of mathematics for you to explore on your own. Everything you need here may be found in 'A little useful mathematics'. Start by drawing an *accurate* diagram. (A good diagram always helps you to think more clearly.)

(2) The chord AP in the circle γ' subtends angles at B and at R. Hence $\angle ABP = \angle ARP$.

(3) The line PR is tangent to the circle γ at R. The angle between the tangent PR and the chord RC is equal to the angle RBC subtended in the opposite segment. Hence $\angle PRA = \angle RBC$.

(4) $\angle ABC = \angle ABP - \angle CBP = \angle RBC - \angle PBC = \angle RBP$

$\qquad = \angle \text{\_\_\_\_\_}$ (angles subtended by the chord RP on γ' are equal).

---

5. An *increasing* sequence of integers is said to be **alternating** if it *starts* with an *odd* term, the second term is *even*, the third term is *odd*, the fourth is *even*. and so on. The empty sequence (with no terms at all!) is considered to be alternating. Let $A(n)$ denote the number of alternating sequences which only involve integers from the set $\{1, 2, \ldots, n\}$. Show that $A(1) = 2$ and $A(2) = 3$. Find the value of $A(20)$, and prove that your value is correct.

---

(1) It is easy to *think* that you have 'solved' a problem like this one, and yet score very few marks. One reason here is that it is tempting just to *guess* the Fibonacci connection, to assume that one's guess is correct, and then calculate as though no further proof is needed. The whole essence of mathematics is that guessing is useful only in that it focuses your attention on *what has to be proved.* You then have to find some way of *proving* that your guess is correct. That is the challenge here. The solution is easy once you find it. But it is *not* obvious.

(2) When $n = 1$, the only alternating sequences are the empty sequence and the sequence '1'; so $A(1) = 2$. When $n = 2$, the only alternating sequences are the empty sequence, the sequence '1', and the sequence '12'; so $A(2) = 3$.

(3) Alternating sequences from $\{1, 2, \ldots, n\}$ are of three types. First, there is the empty sequence; next there are those sequences that begin with a '1' (type A); all other alternating sequences begin with an odd number $\geqslant 3$ (type B).

(4) Each type A sequence arises uniquely by first adding 1 to each term of some alternating sequence from $\{1, 2, \ldots, n-1\}$, and then sticking a '1' at the front. Each type B sequence arises uniquely by adding 2 to each term of some

alternating sequence from $\{1, 2, \ldots, n-2\}$. Since this latter operation also includes the empty sequence, this *proves* that

$$A(n) = A(n-1) + A(n-2).$$

(5) Now use $A(1) = 2$, $A(2) = 3$, and the recurrence relation in (4) to calculate $A(20)$.

29th British Mathematical Olympiad, 1993

1. Find, showing your method, a six-digit integer n with the following properties:
   (a) the number formed by the last three digits of n is exactly one greater than the number formed by the first three digits of n (sc n might look like 123124);
   (b) n is a perfect square.

(1) Look at the condition (a): 'The number formed by the last three digits is one more than the number formed by the first three digits.' What exactly is this saying? Let the number formed by the first three digits be x; then the number formed by the last three digits has to be ____. So if the six-digit number $n = $ '*abcdef*', then $x = $ '*abc*', '*def*' $= x + 1$, and $n = x + 1 + $ ____ . You are also told that n is a square, say $n = y^2$; your equation for n can therefore be written as $y^2 = x + 1 + 10^3 x$.

(2) You should now feel an urge to take the '1' to the LHS and to factorize both sides. (When working with real numbers, this would usually be a bad move. But when working with integers, factorizations can be highly instructive!) Factorize the resulting LHS $y^2 - 1 = ($____$) \cdot ($____$)$, and the RHS $x + 10^3 x = $ ____ $\cdot ($____$)$. Finally, factorize $1001 = $ ____ $\cdot$ ____ $\cdot$ ____ . Hence $(y-1)(y+1) = x \cdot 7 \cdot 11 \cdot 13$.

(3) Now try all possibilities in turn (such as $y - 1 = 7a$, $y + 1 = 143b$, or vice versa). Provided that you are systematic and show your method clearly, you will have anwered the question.

(4) However, the mathematician in you should want to find a deductive approach which will generate *all possible solutions*—such as the following, which is based on the useful fact that $1001 = 7 \cdot 11 \cdot 13$. If $(y-1)(y+1) = 7 \cdot 11 \cdot 13 \cdot x$, then two of the primes 7, 11, and 13 must divide one of factors on

the LHS, while the other prime divides the other factor. (Since $y < 10^3$, at most two of the three primes divide each factor.) There are 3×2 cases:

Case 1 $\quad y + 1 = 143k, \; y - 1 = 7k'$ **Case 2** $\quad y + 1 = 91k, \; y - 1 = 11k'$

$\therefore k = 3, \, k' = 61, \, x = 183$ $\therefore k = \underline{\quad}, \, k' = \underline{\quad}, \, x = \underline{\quad}$

$\therefore y = \underline{\quad}, \, y^2 = \underline{\quad}.$ $\therefore y = \underline{\quad}, \, y^2 = \underline{\quad}.$

Case 1' $\quad y - 1 = 143k, \; y + 1 = 7k'$ **Case 2'** $\quad y - 1 = 91k, \; y + 1 = 11k'$

$\therefore k = \underline{\quad}, \, k' = \underline{\quad}, \, x = \underline{\quad}$ $\therefore k = \underline{\quad}, \, k' = \underline{\quad}, \, x = \underline{\quad}$

$\therefore y = \underline{\quad}, \, y^2 = \underline{\quad}.$ $\therefore y = \underline{\quad}, \, y^2 = \underline{\quad}$ (not six-digit).

Case 3 $\quad y + 1 = 77k, \; y - 1 = 13k'$

$\therefore k = \underline{\quad}, \, k' = \underline{\quad}, \, x = \underline{\quad}$

$\therefore y = \underline{\quad}, \, y^2 = \underline{\quad}.$

Case 3' $\quad y - 1 = 77k, \; y + 1 = 13k'$

$\therefore k = \underline{\quad}, \, k' = \underline{\quad}, \, x = \underline{\quad}$

$\therefore y = \underline{\quad}, \, y^2 = \underline{\quad}$ (not six-digit).

---

2. A square piece of toast *ABCD* of side length 1 and centre *O* is cut in half to form two equal pieces, *ABC* and *CDA*. If the triangle *ABC* has to be cut into two parts of equal area, one would usually cut along the line of symmetry *BO*. However, there are other ways of doing this. Find (with proof!) the length and location of the shortest straight cut which divides the triangle *ABC* into two parts of equal area.

---

(1) Beware of assuming that a line which bisects the area must go through some special point—such as the centroid.

(2) There are *two* essentially different ways of cutting the toast. You may cut off *either* one of the 45° angles (say *A*), or the right angle at *B*. *Both ways have to be considered*—although any method which works in one case ought to work just as well in the other case; thus the second case should be relatively easy.

(3) It is not unreasonable to feel that, because the angle at *A* is *smaller* than the angle at *B*, the shortest cut which cuts off half the triangle by cutting off the angle *A* ought to be shorter than the shortest cut which cuts across the angle *B*, so we consider this first. Let *XY* cut the triangle in half; then area(*ABC*) = $\underline{\qquad}$, so area(*AXY*) = $\underline{\qquad}$. Let $AX = x$ and $AY = y$. Then $\frac{1}{2}$area(*ABC*) = area(*AXY*) = $\frac{1}{2} \cdot xy \cdot 1/\sqrt{2}$.

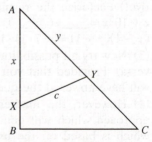

Thus $xy = \underline{\quad}$. Hence as x and y vary over all possible cuts, the product xy remains constant.

(4) How can this help you find c, the length of the cut? It should be clear that you have little choice but to use the $*o*i*e$ rule to find c in terms of x and y, giving $c^2 =$ _____ . You can then substitute for y (using the fixed value of xy) to find an expression for c^2 in terms of x.

(5) To find the shortest cut it is natural (if you know calculus) to differentiate the formula for c^2 with respect to c and solve $dc/dx = 0$. You must then explain clearly why this gives a *minimum*.

(6) If you do not know calculus, or do not wish to use calculus, you should use the constant value of xy to express c^2 in terms of $(x + y)$, and conclude that 'c is a minimum when $x + y$ is a minimum'. To find the minimum value of c for this kind of cut you can then use the AM–GM inequality:

$$\frac{x+y}{2} \geqslant \sqrt{(xy)} \quad \left(= \frac{1}{\sqrt[4]{2}} \right) \qquad \left(\text{with equality if and only if } x = y = \frac{1}{\sqrt[4]{2}} \right).$$

[An even quicker way is to write $c^2 = (x - y)^2 + constant$, and observe that this is a minimum when $x = y$.]

(7) You must now repeat all this for a cut which cuts off the right-angled corner at B, to see if the shortest cut in this case is longer or shorter than the shortest cut in the previous case. The same method works and is left as an exercise.

---

3. For each positive integer c, the sequence u_n of integers is defined by

$$u_1 = 1, \quad u_2 = c, \quad u_n = (2n + 1)u_{n-1} - (n^2 - 1)u_{n-2} \qquad (n \geqslant 3).$$

For which values of c does this sequence have the property that

$$u_i \text{ divides } u_j \text{ whenever } i \leqslant j?$$

---

(1) Faced with a problem like this, there is really only one way of getting started; namely, work out the next few terms of the sequence in terms of c. But be careful! A little algebraic slip at this stage can lead to a lot of misdirected work. So work out the next term u_3 in terms of c: $u_3 =$ _____ .

(2) If you now think about the last sentence of the question, you will see that this tells you something about the two integers u_2 and u_3, and that this in turn tells you that the values of c that you want must satisfy the condition that c divides _____ .

(3) You should now be able to find a shortlist which includes all possible values of c. However, the last sentence of the question is a much stronger condition than just saying 'u_2 divides u_3', so some of these 'possible' values may not work. If c divides a, and c divides b, then c divides the difference $a - b$. You know that c divides $7c$, and you have just shown that c must divide $u_3 =$ _____ ; hence c divides their difference, namely _____ .

(4) This tells you that there are only four possibilities for c (since $c > 0$),

$$c = \underline{\quad}, \text{ or } \underline{\quad}, \text{ or } \underline{\quad}, \text{ or } \underline{\quad}.$$

You can now test each of these possible values in turn (smallest first).

(5) If $c = \underline{\quad}$, the sequence goes $\underline{\quad}, \underline{\quad}, \underline{\quad}, \underline{\quad}, \underline{\quad}, \underline{\quad} = \underline{\quad}$, so u_5 doesn't divide u_6.

(6) If $c = \underline{\quad}$, the sequence goes $\underline{\quad}, \underline{\quad}, \underline{\quad}, \underline{\quad}, \underline{\quad}, \ldots$, which looks promising! How can you *prove* that the sequence has the required property in this case? To have any hope of showing that u_i divides u_j whenever $i \leqslant j$, you really need a *o**u*a for the nth term of the sequence—so try to guess a formula, and then prove it (by i**u**io* on n).

Conjecture. $u_n = \underline{\quad}$.

Proof. The formula certainly works when $n = 1$ (or you would not have chosen it). Thus it will be enough to suppose the formula is valid for the first $n - 1$ terms, and show that it must then be valid for the nth term as well. The definition of u_n says that

$$u_n = (2n + 1)u_{n-1} - (n + 1)(n - 1)u_{n-2}.$$

Since you are assuming that your formula is correct for u_{n-1} and for u_{n-2}, you can substitute for these terms on the RHS and simplify. Do this and complete the proof.

(7) If $c = \underline{\quad}$, the sequence goes $\underline{\quad}, \underline{\quad}, \underline{\quad}, \underline{\quad}, \underline{\quad}, \ldots$, which again looks promising! I leave you to guess and prove a general formula in this case.

(8) Finally, if $c = \underline{\quad}$, the sequence goes $\underline{\quad}, \underline{\quad}, \underline{\quad}, \underline{\quad}, \ldots$, so u_3 does not divide u_4.

---

4. Two circles touch internally at M. A straight line touches the inner circle at P and cuts the outer circle at Q and R. Prove that $\angle QMP = \angle RMP$.

If you have learned a little geometry you should enjoy this problem. The problem is clearly about *angles*, *circles*, and *tangents*. So you expect to have to use the three basic facts about such things; namely,

- the angles in the same segment are equal,
- the opposite angles of a cyclic quadrilateral add to 180°,
- the angle between a chord and a tangent is equal to the angle in
  _____.

(In one sense, the first of these three facts should be enough, since the other two follow from it. However, you should learn to use them all. In this problem the shortest solution avoids using the first fact explicitly.)

(1) Draw your own diagram! So far only four points (namely M, P, Q, and R) and four line segments (namely MQ, MP, MR, and QR) have names. Two other as-yet-unlabelled points are screaming out to be given names. Which points are they?

(2) Let QM and MR meet the inner circle at X and Y respectively. Three line segments (namely ____, ____, and ____) then beg to be drawn in to create lots and lots of ways you can use the three basic facts listed above.

(3) It is tempting now to mark in as many pairs of equal angles as you can. This is a good move; but it can mess up your diagram. So do this on a *rough* diagram, while keeping your best diagram clear until you see which pairs of angles will help you to solve the problem.

(4) From your rough diagram it should soon become clear that, although $\angle QMP$ is equal to several other angles, and $\angle PMR$ is equal to several other angles, there is no direct way of proving that these two angles are equal by simply using our first basic fact. You are clearly going to need the other two facts. The second fact shows

$$\angle QXP = 180° - \angle MXP = \angle \underline{\quad\quad} \qquad \text{(cyclic quadrilateral).}$$

This means that the two triangles QXP and PYM have two angles equal. What about the angles $\angle QPX$ and $\angle PMY$? $\angle QPX$ is the angle between the tangent QP and the chord PX, and so is equal to one of the angles you are interested in: $\angle QPX = \angle \underline{\quad\quad}$ (angles between tangent and chord).

(5) Thus if only you could show that $\angle QPX = \angle PMY$, then you would be finished. One slightly backhanded way of doing this is to show that $\angle PQX = \angle MPY$. To finish things off you have to realize that we have still not used one of the most important hypotheses in the question! Which one? If two circles touch internally at M, then they must have a common $*a**e**$ at M. So you can use our third basic fact to show that the angle between the tangent at M and the chord MR is equal to the very interesting angle subtended by the chord MR at Q on the outer circle—namely $\angle \underline{\quad\quad}$—and the very interesting angle subtended by the chord MY at the point P on the inner circle—namely $\angle \underline{\quad\quad}$.

[Alternatively, you may have noticed two lines in your original diagram which *look* remarkably parallel! It is always worth drawing a careful diagram. And though one should never assume that lines that *look* parallel must be parallel, a diagram can suggest useful things to *prove*. Prove that these two lines are in fact parallel, and use this to solve the problem.]

---

5. Let x, y, and z be positive real numbers satisfying
$$\tfrac{1}{3} \leqslant xy + yz + zx \leqslant 3.$$

Determine the range of values for (a) xyz and (b) $x + y + z$.

(1) The upper bound in (a) is pleasantly easy if you know the AM–GM inequality for three variables, which says that

$$\text{'whenever } a,b,c > 0, \text{ we have } \frac{a+b+c}{3} \geqslant \sqrt[3]{(a \cdot b \cdot c)},$$

$$\text{with equality if and only if } a = b = c\text{'}.$$

To obtain an *upper* bound for xyz, all you need to do is choose a, b, and c in a way which will allow you to use one of the two inequalities involving $xy + yz + zx$ which are given in the question. Once you notice this, there is only one choice, namely $a =$ ____ , $b =$ ____ , and $c =$ ____ . Then

$$1 \geqslant \frac{a+b+c}{3} \geqslant \sqrt[3]{(a \cdot b \cdot c)} \;\Rightarrow\; xyz \leqslant \text{\_\_\_\_},$$

$$\text{with equality when } x = y = z = \text{\_\_\_\_}, \; xyz = \text{\_\_\_\_}.$$

Finally, note that if $x = y = z =$ ____ , then $xy + yz + zx =$ ____ , so the inequalities in the question are satisfied. (To show that you *know* that this upper bound is *exact*, you *must show explicitly* that xyz can actually take this boundary value by specifying explicit values of x, y, and z which satisfy the inequalities given in the question!)

(2) A *lower* bound for xyz may seem much more difficult. Since the upper bound used the right-hand inequality $xy + yz + zx \leqslant 3$, you might expect the lower bound to use the other given inequality, $xy + yz + zx \geqslant \frac{1}{3}$. It doesn't. In fact, xyz can take any positive value $\leqslant 1$. The proof is not hard, but requires a change of focus. Since x, y, and z are all positive, you can be sure that $xyz >$ ____ . To show this is the best possible lower bound, you have to show that it is possible to choose values for x, y, and z which make their product xyz as small as ever you please. To keep things simple, try $x = y = 1$, say. Then z only has to satisfy $\frac{1}{3} \leqslant x^2 + 2xz = 1 + 2z \leqslant 3$; so you can choose z as small as you like and make $xyz = z$ as small as you like. Thus $xyz > 0$ is best possible.

(3) Now for part (b). Once you have understood how to show that xyz can take arbitrarily small positive values, you should be able to show that '$x + y + z$ can be arbitrarily large'. (For example, take $x = 1/n = y$ for $n \geqslant 1$, and choose a suitable value of z such that $\frac{1}{3} \leqslant x^2 + 2xz = (1/n^2) + 2z/n \leqslant 3$: for example, $z = n$.)

(4) The proof of a lower bound for $x + y + z$ is a little more awkward. It helps to notice that $x = y = z = \frac{1}{3}$ gives the value $x + y + z =$ ____ . This seems hard to beat! However, you don't have to spot this before getting started (although you will need it later to show that the lower bound is exact.) Suppose that you want to prove that '$x + y + z \geqslant$ ____ '. You must certainly use the inequalities given in the question. These involve the expression $xy + yz + zx$, so you need to link the expression $xy + yz + zx$ with $x + y + z$ in

some way. One way in which you should certainly be tempted to try is to look at

$$(x + y + z)^2 = x^2 + y^2 + z^2 + 2(xy + yz + zx) \geqslant x^2 + y^2 + z^2 + 2 \cdot (\underline{\quad}).$$

(5) Thus you would like to prove '$x^2 + y^2 + z^2 \geqslant \frac{1}{3}$'. You may either:

(a) use the AM–GM inequality $a^2 + b^2 \geqslant 2ab$ three times to finish the solution; or

(b) rewrite $x^2 + y^2 + z^2 - xy - yz - zx$ as the sum of three squares.

[Alternatively, look up the Cauchy–Schwarz inequality and see if you can use it to produce a shorter solution.]

28th British Mathematical Olympiad, 1992

---

1. (a) Observe that the square of 20 has the same number of non-zero digits as the original number. Does there exist a two-digit number *other than* 10, 20, and 30 the square of which has the same number of non-zero digits as the original number? If you think there is one, then give an example. If you claim that there is none, then you must prove your claim.
   (b) Does there exist a three-digit number *other than* 100, 200, and 300 the square of which has the same number of non-zero digits as the original number?

---

(1) Start with part (a). Let the number be $N = $ 'ab'; that is, $N = 10a + b$. If $b = 0$, then the only solutions are ____, ____, and ____ (since 40, 50, 60, 70, 80, and 90 have squares with two non-zero digits).

(2) Thus you may assume that $b \neq 0$. Hence N^2 has a non-zero *units* digit, and can only have one other non-zero digit. Since the first digit (whether this be the hundreds or the thousands digit) has to be non-zero, *the tens digit must be zero.* Since

$$N^2 = (10a + b)^2 = 10^2 a^2 + 2ab \cdot 10 + b^2,$$

it therefore follows that '$2ab + $ any carry from b^2' must be a multiple of ____; in particular, 'any carry from b^2' must be even. There are very few possibilities:

(i) $b \leqslant 3$, so $10 \mid 2ab$ implies $a = 5$; hence $N = $ ____, ____, or ____: none of these work.

(ii) $b = 5$, which contradicts the fact that $2ab + 2 = 10a + 2$ should be a multiple of 10.

(iii) $b = 7$, and $10 \mid 14a + 4$, so $a =$ _____ $(N =$ _____$)$, or $a =$ _____ $(N =$ _____$)$, neither of which work.

(iv) $b = 8$, and $10 \mid 16a + 6$, so $a =$ _____ $(N =$ _____$)$, or $a =$ _____ $(N =$ _____$)$, neither of which work.

(v) $b = 9$, and $10 \mid 18a + 8$, so $a =$ _____ $(N =$ _____$)$, or $a =$ _____ $(N =$ _____$)$, neither of which work.

Hence $N = 10$, 20, and 30 are the only such two-digit numbers.

(3) Now for part (b). Let the number $N = $ 'abc'; that is, $N = 10^2 a + 10b + c$. If $c = 0$, then $N = 10n$, where n is a two-digit number the square of which has the same number of non-zero digits as n, so the only solutions are $N =$ _____, _____, or _____ (by part (a)). Thus you may assume that $c \neq 0$. If $b = 0$, then $N^2 = 10^4 a^2 + 10^2 \cdot 2ac + c^2$. The first and last digits then have to be non-zero, so all other digits must be zero. Hence $c \leqslant 3$ (as $c \geqslant 4$ gives rise to a carry from c^2, and hence a non-zero tens digit). Moreover, $a \leqslant 3$ (since if $a \geqslant 4$, then $a^2 > 10$, so $10^4 a^2$ would give rise to non-zero digits in both the ten thousands and hundred thousands). But then the hundreds digit will be non-zero. Hence there are no solutions with $b = 0$. Thus we may assume that $b \neq 0$, so

$$N^2 = 10^4 a^2 + 10^3 \cdot 2ab + 10^2(2ac + b^2) + 10 \cdot 2bc + c^2$$

has just three non-zero digits—the first digit, the last digit, *and one other*.

(4) To find an example, it is tempting to try $c = 1$ and to choose $b =$ _____ to kill off the tens digit. Then

$$N^2 = 10^4(a^2 + a) + 10^3 \cdot 2 + 10^2(\text{_____}) + 1,$$

and there are then just two values of a which make the hundreds digit zero; namely, $a =$ _____ or $a =$ _____ . You must then check whether N^2 has just three non-zero digits.

(5) When you think about each of these examples, you should realize that each is closely related to another example. Finding all solutions, with a proof that you have not missed any, requires a little more care; but it is worth doing. (There are fewer than ten examples altogether.)

---

2. Let $ABCDE$ be a pentagon inscribed in a circle. Suppose that AC, BD, CE, DA, and EB are parallel to DE, EA, AB, BC, and CD respectively. Does it follow that the pentagon has to be regular? Justify your claim.

---

(1) You should certainly begin by drawing a pentagon $ABCDE$ inscribed in a circle, with the pentagon *looking* non-regular (to avoid misleading yourself).

(2) To show that the pentagon has to be *regular* you must show that *all five angles are equal* and that *all five sides are equal*. Look first at angles:

- $\angle EAD = \angle EBD = \angle ECD$ (angles in the same segment)
- $\angle EAD = \angle BDA$ (since AE is parallel to BD)
        $= \angle DBC$ (since AD is parallel to BC)
- $\angle BCA = \angle DAC$ (since AD is parallel to BC)
        $= \angle ADE$ (since AC is parallel to ED)
        $= \angle ABE$ (angles in the same segment).

(3) Repeating the reasoning in the final bullet point of (2) once more will show that $\angle ABE = \angle EAD$. It follows that $\angle ABC = 3 \times \angle ABE$. Similarly, $\angle BCD = 3 \times \angle ECD$, and so on, so all five angles of the pentagon $ABCDE$ are equal and each angle is trisected by the diagonals at that corner.

(4) Hence $\angle BAC = \angle BCA$, so $\triangle ABC$ is isosceles, whence $AB = BC$. Similarly, $BC = CD$, and so on, so the five sides of the pentagon $ABCDE$ are all equal. Hence the pentagon is regular.

[Alternatively, there is a lovely short proof: If $\angle ABC = \theta$, then $\angle BCE = 180° - \theta$ (since CE is parallel to AB). But $ABCE$ is a cyclic quadrilateral, so $\angle BAE = \theta$. Continuing in this way, we see that all five angles are equal. It remains to prove that the sides are equal.]

You might like to consider the following variations: (a) What if only four of the diagonals are parallel to the opposite sides? Does it still follow that the pentagon has to be regular? (b) What if you are told not that the pentagon is inscribed in a circle, but that all the sides have the same length? Does it still have to be regular? What if only four of the diagonals are parallel to their opposite sides?

---

3. Find four distinct positive integers the product of which is divisible by the sum of every pair of them. Can you find a set of five or more numbers with the same property?

---

(1) It is natural to begin by trying to construct such a set of four numbers $a < b < c < d$ in the simplest imaginable way: 'If $a = 1$, and $b = 2$, then $a + b = 3$, so you could choose $c = 3$. Then $a + c = 4$ and $b + c = 5$, and $abc = 6$, so $d = 10$ would make $abcd = 60$, which is divisible by the sum of each pair except $a + d$ and $c + d$.'

(2) Various attempts to improve on this should lead you to suspect that $a = 1$ will not work. It is then natural to try $a = $ ___ . 'If $a = 2$ and $b = 3$, then $a + b = $ ___ , so it is natural to try $c = $ ___ . Then $a + c = $ ___ ...'. If this is successful, fine; otherwise, one might try $a = 2$, $b = $ ___ . Provided that you are flexible, imaginative (and persistent) this naive approach should sooner or later produce a set of four numbers which work.

(3) When looking for a set of five numbers, it is natural to try to build on the set of four numbers you have already found. (This strategy is not always the right one in mathematics, but it is always worth trying.) In this case it should work.

(4) At the same time, a little reflection on the set of four numbers which you first stumbled upon should suggest that they are part of a natural sequence $a_1 < a_2 < a_3 < \ldots$. If you produce such a sequence and *prove* that it has the property that

'for every $m \geqslant 4$, $a_1 a_2 a_3 \ldots a_m$ is divisible by each sum $a_i + a_j$ $(1 \leqslant i \neq j \leqslant m)$',

you will have answered the question as fully as required. (Let $a_i = 4i - 2$. Suppose that $i < j$. Then $a_i + a_j = 4(i + j - 1)$. If $i + j$ is odd, then we may write $4(i + j - 1) = 2^x k$ with k odd; now $k \leqslant (i + j - 1)/2 < j$, so $k \mid 1 \cdot 3 \cdot 5 \cdot \ldots \cdot (2j - 1)$, whence $2^x k \mid a_1 a_2 \ldots a_j$ (since $x \leqslant j$). If $i + j$ is even, then $i + j - 1 = 2m - 1$ is odd with $m < j$, so $a_i + a_j = 2a_m \mid a_1 a_2 \ldots a_j$.)

(5) There remains the (unstated) question as to whether the sequence $\{a_i\}$ in (4) is the only sequence of this kind.

---

4. Determine the smallest value of $x^2 + 5y^2 + 8z^2$, where x, y, and z are real numbers subject to the condition $yz + zx + xy = -1$. Does $x^2 + 5y^2 + 8z^2$ have a *greatest* value subject to the same condition? Justify your claim.

Problems of this kind can seem impossibly difficult at first. However, when you eventually see how to do them you are likely to kick yourself, for at this level many inequalities are simply disguised versions of the basic fact that *squares cannot be negative*.

(1) The given expression consists of squares with positive coefficients and so can never be negative. But it is not enough just to say that it is always $\geqslant 0$, for the expression can never actually equal 0 ($x = y = z = 0$ does not satisfy the constraint '$xy + yz + zx = -1$').

(2) Is there some way in which you can write the expression $x^2 + 5y^2 + 8z^2$ in terms of perfect squares which will bring in terms involving xy and yz, and zx? Of course there is! You expect terms such as xy to appear whenever the bracket you are squaring contains an x term and a y term. So the vague question above can now be sharpened up in the following way. Can you rewrite the given expression $x^2 + 5y^2 + 8z^2$ as a sum of squares of two or more separate brackets in such a way that the brackets between them contribute $1 \cdot x^2$, $5 \cdot y^2$, and $8 \cdot z^2$, plus mixed terms which can be simplified using $xy + yz + zx = -1$? (Once you have done this you should obtain an obvious lower bound just by observing that the squares have to be $\geqslant 0$. However, you should not stop there. How do you know that the lower bound

that you have found is the best possible one—corresponding to the 'smallest possible value' of the given expression? The answer is that you don't, unless you manage to find *specific* values of x, y, and z (a) which make each of the squared brackets equal to zero, and (b) which also satisfy the constraint $xy + yz + zx = -1$.)

(3) The first thing to notice is that $(ax + \ldots)^2$ contributes $a^2 x^2$. Since we may assume that $a \geqslant 0$, to land up with $1 \cdot x^2$ you have to have $a = 1$ with $1 \cdot x$ in just one bracket. And to land up with $5 \cdot y^2$ you have to have $(\pm 1) \cdot y$ in one bracket and $(\pm 2) \cdot y$ in another. And to land up with $8 \cdot z^2$ you have to have $(\pm 2) \cdot z$ in one bracket and $(\pm 2) \cdot z$ in another. This suggests just two brackets with relatively few possibilities:

$$\left(x \pm \left\{ \begin{matrix} y \\ 2y \end{matrix} \right\} \pm 2z \right)^2 + \left(\left\{ \begin{matrix} y \\ 2y \end{matrix} \right\} \pm 2z \right)^2 .$$

(4) You have to choose the signs (and which y term goes in which bracket) to try to ensure that the mixed terms involving xy, yz, and zx all have the same coefficient. This coefficient turns out to be -4, so the constraint shows that

$$(*) \quad \begin{array}{cccccc} x^2 + 5y^2 + 8z^2 & = & (x + 2y + 2z)^2 & + & (y - 2z)^2 & - & 4(xy + yz + zx) \\ & \geqslant & 0 & + & 0 & + & 4 \end{array}.$$

(5) If you want to choose x, y, and z actually to achieve the value '4', you must first make the two squares on the RHS of $(*)$ equal to 0 by insisting that $y = 2z$ and that $x = -2y - 2z = -3y$. You must then fix up these values of x, y, and z to satisfy $xy + yz + zx = -1$. For example, if you try $z = 1$, $y = \underline{}$, and $x = \underline{}$ to satisfy the condition, then $xy + yz + zx = \underline{}$; hence to satisfy the correct constraint while keeping the brackets equal to 0 it is enough to divide each of the trial values of x, y, and z by $\underline{}$. Hence the expression $x^2 + 5y^2 + 8z^2$, subject to the constraint $xy + yz + zx = -1$, really does attain its smallest value 4.

(6) How about a greatest value? You may be inclined to suspect that the answer in this case may be 'No', and to try to show that the expression can take arbitrarily large values. If you decide to try this, it is important to simplify as much as possible. Could you perhaps get rid of one of the three variables? For example, if $z = 0$, then the constraint becomes $x = -(1/y)$. If you then let y become larger and larger, x tends to zero; thus the given expression $x^2 + 5y^2 + 8z^2$ just goes on getting larger and larger without bound. Hence it has no greatest value.

---

5. Let f be a function mapping positive integers into positive integers. Suppose that

$$f(n + 1) > f(n) \quad \text{and} \quad f(f(n)) = 3n \quad \text{for all positive integers } n.$$

Determine $f(1992)$.

(1) Successful problem-solving often depends on clinging to the optimistic assumption that, although unfamiliar problems often seem impossible at first, they can usually be solved once you understand what is going on. This question looks unfamiliar, and is at first sight *very abstract*—in that it involves an unknown function f which is not actually given, being described only in terms of some of its properties. Despite all this, you have no option but to take courage and set to! Where should one begin? The obvious place (since f maps positive integers to positive integers) is to start thinking about the value of $f(n)$ when $n =$ ____ . Since $f(1)$ has to be a *positive* integer, you certainly know that $f(1) \geqslant 1$. You also know that $f(f(1)) =$ ____ . It follows that f cannot be the identity function. This may not seem to help, until you realize that this, combined with the condition $f(n + 1) > f(n)$, implies that $f(n) > n$ (Why?). Use this fact to prove that $f(1) = 2$.

(2) Use the same ideas to find $f(2)$, $f(3)$, and $f(6)$.

(3) Once you know $f(3)$ and $f(6)$, the condition $f(n + 1) > f(n)$ should tell you exactly what $f(4)$ and $f(5)$ have to be.

(4) You are now in business, since the condition $f(f(n)) = 3n$ allows you to fill in quite a few other values (such as $f(7)$, $f(8)$, $f(9)$, and so on). It may still not be clear what is going on, but at least the horrible feeling that you cannot even begin should have begun to recede.

(5) If you construct a table of values for the function, you should realize that there are long stretches where $f(n + 1) = f(n) + 1$, followed by stretches where the value of $f(n)$ goes up in jumps. How large are these jumps? Later on you will have to come back and prove the important bits, but the main thing at this stage is to convince yourself that what seemed like a totally inaccessible problem is in fact much more manageable than you thought. The first step in making sense of what is going on is often to risk yourself by making some kind of a *guess* as to what you expect to find in the next bit of the table—that is, to formulate simple *conjectures*, and then to test them against the facts to see if they stand up. If your guess turns out to be wrong, stand back and try to improve it. You should decide fairly quickly that you think you know where the jumps occur. It is not too hard then to decide what you *expect* the value of $f(1992)$ to be.

(6) All that then remains is to decide what to prove, and how to prove it. This is not nearly as hard as it may appear provided that you sort out what to prove first:

(a) Prove first (by induction on n) that $f(3^n) =$ ____ and that $f(2 \cdot 3^n) =$ ____ .

(b) Then prove that $f(3^n + k) =$ _____ for each k, $0 \leqslant k \leqslant 3^n$.

(c) Finally, deduce that $f(2 \cdot 3^n + k) = 3^{n+1} + 3k$ for each k, $0 \leqslant k \leqslant 3^n$.

You can then use the proven facts (a), (b), and (c) to calculate $f(2 \cdot 3^6) =$ ____ , and $f(2 \cdot 3^6 + 534) =$ ____ .

27th British Mathematical Olympiad, 1991

---

1. Prove that the number $3^n + 2 \times 17^n$, where n is a non-negative integer, is never a perfect square.

---

(1) Let $u_n = 3^n + 2 \times 17^n$. Work out the first four or five terms, keeping a look-out for anything which might be helpful. (Of course, you could simply look at the numbers that you obtain for each of the first few terms and see whether they happen to be perfect squares. But that would miss the point! The challenge here is to *prove* that u_n can *never* be a perfect square *for any value of n*. So what you should be looking for is something about the values you are getting which has a chance of being true *for every* u_n, and which might give you a clue as to why u_n can *never* be a square.)

(2) As n increases it takes longer to calculate u_n (and the risk of making an error increases!). This makes it important not to waste information. What was the first value of n you should have used in (1)? (*Hint:* the answer is *not* $n = 1$.)

(3) Look carefully at the first few terms u_n. Do you notice anything which is incompatible with being a perfect square and which might be true for later terms as well? (In many mathematical problems it can be a disadvantage to know too much. Fortunately, there are only two basic facts about perfect squares, and most people know only one of them! So you are unlikely to be handicapped here by knowing too much.)

(4) Write out the first 12 perfect squares and compare your list with the numbers $u_0, u_1, u_2, \ldots$. What differences stand out? This should give you *an idea*. However, the crucial word in Question 1 is the first word: 'Prove'. It is not enough just to *assert* that the units digit of u_n follows some pattern (that would get you at most 1 mark out of 4!). Patterns can deceive. It is your job not just to assert, but to *prove* that what you think really happens does indeed go on for ever.

(5) Any problem in which you have to prove that something is true *for every integer $n \geqslant 0$* should immediately suggest the method of *proof by induction*. This leaves *you* to decide exactly what to prove and how to prove it. (Almost any approach should convince you that it is much easier to work with the two parts 3^n and 2×17^n *separately*.) You should formulate, and prove by induction, a statement about the units digit of any number of the form 3^n ($n \geqslant 0$). Do the same for 2×17^n. Hence solve Question 1.

[You might like to try the following variation. For which values of n could $4^n + 3 \times 18^n$ conceivably by a perfect square? Examine these values of n more carefully and hence decide (and prove your assertion) whether it can in fact ever be a perfect square.]

2. Find all positive integers k such that the polynomial $x^{2k+1} + x + 1$ is divisible by the polynomial $x^k + x + 1$. For each such k, specify the integers n such that $x^n + x + 1$ is divisible by $x^k + x + 1$.

(1) The first part of the question simply asks when $x^{2k+1} + x + 1$ is divisible by $x^k + x + 1$. The natural approach is therefore to try to carry out the long division and see what happens. Do this, keeping track of the remainder as you go along. (The difference between this and ordinary long division of polynomials is that here you are *not* trying to find the answer! Instead, you want to discover which values of k could conceivably leave no remainder.)

(2) At some point in the division you should get a non-zero remainder $r(x)$ which does not involve k. Find $r(x)$. If $x^{k+1} + x + 1$ is to divide $x^{2k+1} + x + 1$ exactly, with no remainder, then $x^k + x + 1$ must divide this remainder $r(x)$ exactly. Hence the degree of $x^k + x + 1$ must be less than or equal to the degree of the polynomial $r(x)$. Now check each of the four possible values of k to see which, if any, of them actually works.

(3) In the second part of the question, the appearance of three letters x, n, and k can easily mislead you. Things are not nearly as complicated as they seem, since the letter k merely stands for the value you found in the first part! Rewrite the second part of the question, replacing the letter k by its actual value. Then carry out the long division of $x^n + x + 1$ by $x^k + x + 1$ and find all values of n for which the division can end with remainder equal to zero.

(4) Whether or not your first approach actually works, it is always worth looking back to see if there is some way of improving, or simplifying, what you did. How can you be sure *at a glance* that 7 divides 6993 exactly *without actually doing the division*? (The answer has something to do with $6993 + 7$.) A polynomial $f(x)$ divides exactly into another polynomial $g(x)$ precisely when $f(x)$ also divides the *sum* $g(x) + f(x)$; similarly, $f(x)$ divides $g(x)$ precisely when $f(x)$ divides $g(x) - f(x)$. Use this idea with $f(x) = x^k + x + 1$, $g(x) = x^{2k+1} + x + 1$. How does this simplify the problem? (You may still feel that you prefer the original approach. However, this idea of manipulating polynomials like numbers is very powerful. At some stage in any mathematical problem you have to know something, or work it out on the spot, or be prepared to go away and learn it!

- Factorize $x^{2k+1} - x^k$ in the obvious way.
- Show that x^k and $x^k + x + 1$ can have no common factors. Hence conclude that $x^k + x + 1$ divides $x^{2k+1} + x + 1$ precisely when $x^k + x + 1$ divides $x^{k+1} - 1$.
- Hence solve the first part of Question 2.
- Use the same idea (or put $x = \exp(2\pi i/3)$ and use the Remainder Theorem) to find all values of n for which $x^n + x + 1$ is divisible by $x^2 + x + 1$ (without remainder).)

3. *ABCD* is a quadrilateral inscribed in a circle of radius *r*. The diagonals *AC* and *BD* meet at *E*. Prove that if *AC* is perpendicular to *BD*, then

(∗) $EA^2 + EB^2 + EC^2 + ED^2 = 4r^2.$

Is it true that if (∗) holds then *AC* is perpendicular to *BD*? Give a reason for your answer.

(1) In any geometry question, you *must* begin by drawing a suitable diagram.
(2) All those right angles (and the *squares* in the equation (∗) that you are trying to prove) should suggest that you need ∗∗∗∗a∗o∗a∗' theorem. Use this theorem to write down as many equations as you can. Choose two of these equations and add to obtain an equation of the form

$$EA^2 + EB^2 + EC^2 + ED^2 = \ldots(**).$$

(3) The RHS of your equation (∗∗) involves terms such as AB^2 which do *not* appear in (∗), and does not yet involve the crucial quantity r^2. The need to involve r^2 should give you itchy fingers, so that you can scarcely resist the temptation to mark in the centre *O* of the circle in your diagram and then to draw in the four obvious radii.
(4) This does not finish off the problem, but it should begin to get you excited. For there are now four new triangles staring you in the face, each with two sides of length *r* and with one other side which appeared in one of the (many!) equations that you wrote down in (2). Use the ∗o∗i∗e rule (on triangles *AOB* and *COD* or on triangles *BOC* and *DOA*) to substitute for $AB^2 + CD^2$ (or for $BC^2 + DA^2$) in terms of *r*.
(5) Show that the equation (∗) is true if $\cos\angle AOB + \cos\angle COD = 0$ (or $\cos\angle BOC + \cos\angle DOA = 0$); that is, if $\angle AOB + \angle COD = 180°$. Then use the fact that *AC* and *BD* are perpendicular to complete your solution to the first part of the question.
   [Alternatively, you may prefer a solution to the first part which avoids the cosine rule:

$$EA^2 + EB^2 = AB^2 \qquad \text{and} \qquad EC^2 + ED^2 = CD^2.$$

So we only have to show that $AB^2 + CD^2 = (2r)^2$. Let *BF* be the diameter through *B*. Then

$\angle AFB = \angle ADB$	(angles in the same segment)
$\angle FAB = 90°$	(angle in a semi-circle)
$\angle DEA = 90°$	(given)
$\therefore \ \angle ABF = 90° - \angle AFB = 90° - \angle ADE = \angle DAE.$	
$\therefore \qquad AF = CD$	(chords subtending equal angles)
$\therefore \quad (2r)^2 = AB^2 + AF^2$	(since $\triangle ABF$ is right-angled)
$\qquad = AB^2 + CD^2.$]	

(6) For the second part of the question you must think carefully about the method that you have just used to answer the first part. The two sides of equation (∗) turned out to be equal because, *when AC is perpendicular to BD*, both sides are equal to $AB^2 + CD^2$ (or $BC^2 + DA^2$). (For the LHS we used Pythagoras (a special case of the cosine rule) on the right-angled triangles *ABE* and *CDE* meeting at *E*, while for the RHS we used the cosine rule on the triangles *AOB* and *COD* meeting at *O*.) You could do exactly the same for any quadrilateral even when *AC* is not perpendicular to *BD*: all you need for (∗) to hold is that cosine terms which arise at *E* should cancel with the cosine terms which arise at *O*. When the diagonals *AC* and *BD* are perpendicular, the cosine terms on each side sum to zero and so balance automatically. Another easy way of making the cosine terms on each side balance is if the diagonals happen to cross at the point ___ (that is, when *ABCD* is a ∗e∗∗a∗∗∗e). Hence the converse is ∗a∗∗e.

[You might like to find out what one can say about an arbitrary cyclic quadrilateral *ABCD* if all you know about it is that it satisfies the condition (∗).]

---

4. Find, with proof, the minimum value of $(x + y)(y + z)$, where x, y, and z are positive real numbers satisfying the condition $xyz(x + y + z) = 1$.

---

(1) This question may look quite unlike any sort of problem you are familiar with. But that does not mean that it is beyond you! Good mathematical problems are often like this. When you first read them you are left totally bemused. But as you struggle to find a way to begin, and as you try out one or two ideas, you gradually realize that they may not be all that hard. If you then stick at it, a complete solution should emerge sooner rather than later. (After experiencing this initial-frustration-leading-to-eventual-success a few times, you will begin to realize that you should never be put off by first impressions.) One off-putting feature of Question 4 is that there are *three* unknowns, x, y, and z. If there were only two unknowns (x and y) you might interpret a 'condition' involving x and y as the equation of a curve in two dimensions. You could then sketch the curve (geometry), or solve the equation to find y in terms of x (algebra). The appearance of three unknowns makes the question *appear* more forbidding. Admittedly the geometry is a little more difficult: the equation $xyz(x + y + z) = 1$ is the equation of a surface in the positive 'octant' ($x > 0$, $y > 0$, $z > 0$)—and we all find 3-D geometry harder to visualize than 2-D geometry. But, as Descartes showed when he invented co-ordinate geometry, algebraic fools can rush in *and succeed* where geometric angels fear to tread! Algebra often generalizes easily, and can be applied routinely; whereas geometry requires real thought and insight.

(2) As the equation $xyz(x + y + z) = 1$ stands, it is difficult to relate it to the expression $(x + y)(y + z)$ which we are trying to minimize. It would be much

easier if the two could somehow be combined, so that the condition $xyz(x + y + z) = 1$ becomes *incorporated in* the expression $(x + y)(y + z)$: you could then concentrate on one expression and forget about the extraneous condition. The most obvious way of incorporating the condition $xyz(x + y + z) = 1$ is to use it to *u***i*u*e for one of the variables in the expression $(x + y)(y + z)$.

(3) But which variable should you eliminate? The expression $(x + y)(y + z)$ is only partly symmetrical in x, y, and z. Which one of the three variables in the expression $(x + y)(y + z)$ sticks out like a sore thumb? Use the equation $xyz(x + y + z) = 1$ to express this variable in terms of the other two. Then substitute this back in the expression $(x + y)(y + z)$ and simplify.

(4) This does not solve the problem. But it certainly looks like progress in that the expression you have to minimize now has only two variables and is beautifully symmetrical. So you no longer need to be so frightened by it! You might even try substituting a few easy values for x and z to see what values the expression takes. (The fact that you have substituted for y using the condition $xyz(x + y + z) = 1$ means that this condition is now automatically satisfied.) For example, find the value of y when $x = z = 1$. What is the value of the expression $(x + y)(y + z)$ in this case?

(5) To finish things off properly you must now prove (using the AM–GM inequality for two variables, or otherwise) that for any positive real number u, $u + (1/u) \geqslant 2$. Hence complete your solution of Question 4.

(6) Having struggled to produce a solution, it is always worth looking back over it to see if there is anything you can learn from it. You should now realize the importance of incorporating the condition $xyz(x + y + z) = 1$ into the expression $(x + y)(y + z)$ in some way. Can you see another way to rewrite $(x + y)(y + z)$ which will allow you to substitute *very simply* using the fact that $xyz(x + y + z) = 1$, and which leads very easily to the same simplified expression as in (3)?

---

5. Find the number of permutations (arrangements)

$$j_1, j_2, j_3, j_4, j_5, j_6 \quad \text{of} \quad 1, 2, 3, 4, 5, 6$$

with the property

for no integer n, $1 \leqslant n \leqslant 5$, do $j_1, j_2, \ldots, j_n$

form a permutation of $1, 2, \ldots, n$.

---

(1) At a first reading it may not even be clear what the question is saying. Do the symbols $j_1, j_2, j_3, j_4, j_5,$ and j_6 stand for six different permutations of 1, 2, 3, 4, 5, and 6? Or is '$j_1, j_2, j_3, j_4, j_5, j_6$' a single permutation of '1, 2, 3, 4, 5, 6'? One of these two interpretations makes a nonsense of the property at the end of the question and so is obviously wrong! So which is correct?

(2) Once you have sorted out what the words mean, you may still not see how to begin. In that case you could begin to work your way into the question by trying to calculate the corresponding number of permutations for some easier cases. Let p_n be the number of permutations $j_1, j_2, \ldots, j_n$ of $1, 2, \ldots, n$ with the property:

($*$)   for no $k < n$ is $j_1, j_2, \ldots, j_k$ a permutation of $1, 2, \ldots, k$.

Clearly, $p_1 = 1$. Work out p_2. Then work out p_3.

(3) You should not expect to find some simple-minded pattern: mathematics is much more interesting than that! It is easy to get the wrong idea about why it is often worth looking at what happens for small values of n. The answers themselves are not very interesting. Instead, you should be looking for a manageable *method* which might help you to calculate p_6—a number which in this case may well be around 500, since the number of all permutations of 1, 2, 3, 4, 5, and 6 is $6! = \underline{\quad}$. (If you try to count without a suitable method, you are bound to make mistakes—probably already when calculating p_4, and almost certainly when calculating p_5 and p_6.) Try to think of a general method for calculating p_4. Use your method to find p_4. Finally, write down (in some systematic way) *all* permutations j_1, j_2, j_3, j_4 of 1, 2, 3, 4 satisfying the condition ($*$) in (2) above, count them, and so check your value for p_4.

(4) There are many possible general methods. Perhaps the most simple-minded one is to observe that p_4 counts

all permutations j_1, j_2, j_3, j_4 of $1, 2, 3, 4$,

except those which do *not* satisfy ($*$).

(a) How many permutations j_1, j_2, j_3, j_4 of 1, 2, 3, 4 have $j_1 = 1$? (b) How many have $\{j_1, j_2\} = \{1, 2\}$? (c) How many have $\{j_1, j_2, j_3\} = \{1, 2, 3\}$? What value does this suggest for p_4?

(5) Unless you are alert, the values you obtained for p_4 in (3) and in (4) will be different! What can have gone wrong? The method in (4)(a) counts all permutations of the form 1, __, __, __; whereas the method in (4)(b) counts all permutations of the form 1, 2, __, __ or 2, 1, __, __. Clearly, some permutations get counted in *both* (4)(a) and (4)(b) (and again in (4)(c)). To avoid this, (4)(b)—for example—should be adjusted so that it only counts permutations of the form 2, 1, __, __, (that is, permutations j_1, j_2, j_3, j_4 of 1, 2, 3, 4 such that $\{j_1, j_2\} = \{1, 2\}$ *but* $j_1 \neq 1$). How many of these are there?

(6) Show that $p_4 = 4! - (p_1 \times 3! + p_2 \times 2! + p_3 \times 1!)$. Find p_5. Then find p_6.
[You might like to try to find a formula for p_n.]

6. Show that if x and y are positive integers such that $x^2 + y^2 - x$ is divisible by $2xy$, then x is a perfect square.

(1) Like many good mathematics problems, this one looks a little strange at first sight. Working your way into a problem can take a long time. You have to show that, under the given conditions, the positive integer x *has* to be a square. There are clever ways of doing this, but, as so often with Olympiad problems, there is also a very simple-minded approach—provided that you have done your homework on 'A little useful mathematics'. How can one recognize when x is a perfect square? One way is to imagine the number x already factorized as a product of prime powers $x = p_1^u \cdot p_2^v \cdot \ldots \cdot p_m^w$. Then x will be a perfect square precisely when all the exponents $u, v, \ldots, w$ are e*e*.

(2) Suppose that p is a prime number which divides x. Show that p must also divide y. Hence show that p^2 must also divide x.

(3) This certainly looks promising, but we need a more general observation for a complete solution. Suppose that p is a prime number such that some odd power p^{2i-1} divides x. Show that then p^{2i} must divide x. Hence write out a complete solution to Question 6.

[Alternatively, you may interpret the given condition '$x^2 + y^2 - x$ is divisible by $2xy$' as an equation; solve it (as a quadratic in _____); hence find a different solution to Question 6.]

---

7. A ladder of length l rests against a vertical wall. Suppose that there is a rung on the ladder which has the same distance d from both the wall and the (horizontal) ground. Find *explicitly*, in terms of l and d, the height h from the ground that the ladder reaches up the wall.

---

(1) Draw a diagram and mark the lengths l, d, and h.

(2) It is natural to try to solve the problem *directly*. Let the foot of the ladder be distance f from the wall. Use *i*i*a* triangles to show that $f/h = (f-d)/d$. Hence find f in terms of h and d. Then use Pythagoras' Theorem to obtain an equation involving only h, d, and l. Hence try to find an expression for h in terms of l and d.

(3) The difficulty with this straightforward approach is that the equation that one obtains linking h, l, and d is a *quartic* in h (that is, it involves h^4), and you are unlikely to know a formula for the roots of such an equation. You would obviously *prefer* a *ua**a*i* equation—that is, one involving only h and h^2 (as well as l and d). One possible way of finding such an equation for h is to notice that, if h does indeed satisfy a quadratic, then the quartic you obtained in (2) must factorize as the product of two quadratics. So it is natural to try to factorize the quartic that you found in (2).

(4) While this must be possible, there is no reason why the coefficient of h^2, the coefficient of h, and the constant term in each quadratic factor should be simple monomials (such as ld, or l^2): they might, for example, involve square roots. This makes the task of factorizing the quartic rather hard. It may be easier to look for a *direct* way of producing a quadratic. For that you need an

idea which explains why the length h should satisfy a quadratic. Look again at the diagram that you drew in (1). Any equation which has h as one root will automatically have another root corresponding to one of the other lengths in your diagram. Which length is it? Why must the roots come in pairs like this? (At first sight you may think there is an *asymmetry* between the vertical height h up the wall and the horizontal distance f from the wall to the foot of the ladder. But if you turn the diagram through a right angle it should be clear that *geometrically* the two lengths h and f are indistinguishable. Thus one should expect that any equation which has h as one root will have f as another root and that alebraically there will be no way of distinguishing between them.)

(5) If the roots of a quadratic $x^2 - px + q$ are α and β, what are p and q? Thus all you need to do is to find expressions for $h + f$ and for $h \cdot f$ in terms of l and d only. (*Hint:* If $h^2 - ph + q = 0$, show that $p^2 = l^2 + 2q$ and $dp = q$. Hence find p and q.)

26th Britich Mathematical Olympiad, 1990

> 1. Find a positive integer the first digit of which is 1 and which has the property that, if this digit is transferred to the end of the number, the number is tripled.

(1) Which is the 'first digit'? And what is meant by 'transferred to the end of the number'? A moment's thought should make it clear that the first digit must be the one on the *left-hand end* and that 'transferred to the end' must mean 'moved to the *right-hand end*'. (Whenever you are faced with a slightly unusual problem, it is tempting to think that the question is unclear in some way. But a more careful reading will nearly always show that there is only one possible interpretation. In this case, if the 1 were moved to the *left*-hand end, the final number could not be even *twice* as large as the original number! Thus the original number must start out with a 1 on the left-hand end, and the 1 must get moved to the right-hand end.)

(2) Since we do not know how long the original number is, we may as well write it as 1_____. We must then solve either

$$\begin{array}{r} 1\_\_\_\_ \\ \underline{3 \times} \\ \underline{\_\_\_\_ 1} \end{array} \qquad \text{or} \qquad 3\overline{)1\_\_\_\_1}$$

So it is really a question about multiplication (or division). Which operation is easier to think about—multiplication, or division?

(3) Concentrate on the first version—the easier one. The only unusual thing

about this is that we are given the multiplier '3', and the units digit '1' of the answer, and we have to work out the units digit of the *original number*:

- What must the units digit of the original number be?
- Which other previously unknown digit can you now fill in?
- What does this now tell you about the tens digit of the original number?
- Keep going.

(4) Of course it is important to remember that the two extended blanks '_____' in (2) represent exactly the *same string of digits*. So once you know the units digit of the original number, you immediately know the ✶e✶✶ digit of the answer. Knowing this and the multiplier '3' then allows you to work out what the tens digit of the original number must be. But this is the same as the ✶u✶✶✶e✶✶ digit of the answer—and so on.

(5) How do you know when to stop? (How many different answers are there?)

(6) Solving a problem is only the beginning. When you come to write out your final solution you should always look for a more mathematical approach. The method used above is easy to carry out, but not so easy to explain in words. One reason for this difficulty is that you have found a *procedure* that works, and this tends to block further thought about *why* it works. A non-mathematician may be content to have a method that works; but the mathematician always wants to know what makes a procedure work. One approach here is to denote the extended blank '_____' in the original number by 'x':

- The blank in the answer may be the same string of digits; but they have been shifted one place to the left. Thus the blank in the answer is not x, but _____ .
- If x has n digits, then the 1 at the front denotes not 1, but _____ , so the original number was $x +$ _____ .

(7) The question now gives rise to an equation involving x and n:

- What is this equation?
- How can you solve it to find (n and) x? How many possible answers are there?
- What is the connection between these answers and the (recurring) decimal for $\frac{3}{7}$?

[You might like to try the following variation. Find a positive integer the first digit of which is a and which has the property that, if this digit is transferred to the end of the number, the resulting number is exactly b times the original:

(a) For which integer values of a and b does this problem have a solution?

(b) Give a complete description of all possible solutions for each pair a, b.]

---

2. $ABCD$ is a square and P is a point on the line AB. Find the maximum and minimum values of the ratio PC/PD, showing that these occur for the points P given by $AP \times BP = AB^2$.

---

(1) If P is a point on the line AB, you may think that PC/PD is a maximum when $P = A$, and that PC/PD is a minimum when $P = B$. Why is this wrong?
(2) As always, *you* have to sort out what the question really means: that is part of the challenge. In this case it clearly depends on what is meant by 'the line AB'. Even if you jump to the conclusion and (mis)interpret this as meaning 'the line segment *between* A and B', you should notice two things. First, if $P = A$, then the required condition $AP \times BP = AB^2$ does not hold, so something is clearly wrong. Second, the ratio PC/PD clearly increases as P moves from B towards A, and so presumably goes on increasing for a while if P continues to move *past A* on the line BA. (Could it increase for ever?) All this suggests that 'the line AB' must mean the *infinite* line through A and B.
(3) Suppose that A is to the left of B. Imagine P moving along the infinite line AB:

- What happens to the ratio PC/PD as P whizzes off past B to $+\infty$.
- What happens to the ratio PC/PD as P whizzes off past A to $-\infty$.

If A is the top left-hand corner of the square $ABCD$, then the point $P = P_{max}$ at which PC/PD is a *maximum* must occur when P is somewhere to the *left* of A. The point $P' = P_{min}$, at which $P'C/P'D$ is a *minimum*, occurs when $P'D/P'C$ is a *maximum*; hence P_{min} must be exactly as far to the *right of B* as P_{max} was to the left of A.
(4) Now that you know what you are looking for, it is natural to take a general point P and call something 'x'; then use Pythagoras to express PC and PD in terms of x, and find an expression for PC/PD in terms of x; finally, find the value of x at which this expression is a maximum or a minimum, and then solve the original problem.
(5) There are many other approaches. For example, the answer '$AP \times BP = AB^2$' is given! And, while products occur relatively rarely in geometry, one often comes across $*a*io*$—especially in triangles. Thus, the given relation $AP \times BP = AB^2$ should suggest looking at PA/AB (which is equal to PA/AD —that is, $\tan \angle PDA$). Let $\angle PDA = \theta$. Find PD in terms of θ and the side s of the square $ABCD$ (using $\triangle PAD$). Find $(PC/PD)^2$ in terms of θ (using the cosine rule in $\triangle PDC$). Show that $(PC/PD)^2 = 1 + \cos^2\theta + \sin 2\theta$. Conclude that, for PC/PD to be a maximum or a minimum, $t = \tan \theta$ must satisfy $t^2 + t - 1 = 0$. Hence complete the solution.

[You might like to try the following variation. Let C and D be any two points in the plane and let $k > 0$ be any real number:

(a) Show that the locus of points P such that $PC/PD = k$ is (i) a circle with C inside if $k < 1$; (ii) the perpendicular bisector of the segment CD if $k = 1$; (iii) a circle with D inside if $k > 1$.

(b) Show that the point P on the line AB at which PC/PD is a minimum corresponds to the value of k for which the circle in part (a) just touches the line AB.]

---

3. The angles A, B, C, and D of a convex quadrilateral satisfy the relation

$$\cos A + \cos B + \cos C + \cos D = 0.$$

Prove that $ABCD$ is either a trapezium or is cyclic.

---

(1) A *convex* quadrilateral is a quadrilateral with no 'dents'—that is, with all angles $\leqslant 180°$. It is often a good idea to begin by writing down the *simplest* facts which seem to be relevant. They may not be the most important, but one should certainly not overlook them. (Moreover, putting *something* down on paper can help to get things moving and give one a feeling of progress.) What do you know about the four angles A, B, C, and D in any quadrilateral $ABCD$? Use this fact to express $\cos D$ in terms of A, B, and C.

(2) So, if you wish, you can eliminate D from the given equation $\cos A + \cos B + \cos C + \cos D = 0$. But that would spoil the symmetry, so perhaps you should stay with the original equation for a while. What do you know about the angles A and C (or B and D) of a cyclic quadrilateral $ABCD$? In what sense do the angles A, B, C, and D in a trapezium $ABCD$ satisfy something similar?

(3) Your answers to the two questions in (2) suggest that, whenever a convex quadrilateral $ABCD$ satisfies $\cos A + \cos B + \cos C + \cos D = 0$, you should try to prove that either two opposite angles, or two adjacent angles, have sum equal to ____. It is time to go back to the original question. What could you possibly *do* with a sum of four cosines? (You really have very little choice, for there is only one obvious thing to do with a sum of cosines!)

(4) But even if you had to make a choice, you should realize that, while one can never conclude very much from an equation of the form 'sum of things $= 0$', one can conclude much more from an equation of the form '*product* of things $= 0$'. So you should instinctively *want* to change the *sum* on the LHS of the original equation into a *product*:

- What is the trig identity that expresses $\cos X + \cos Y$ as a *product* of cosines?

- Apply this to $\cos A + \cos B$ and to $\cos C + \cos D$.
- Factorize the expression for $\cos A + \cos B + \cos C + \cos D$. (Why is $\cos \frac{1}{2}(A + B) = -\cos \frac{1}{2}(C + D)$?)
- Use the identity for $\cos X - \cos Y$ to express the sum $\cos A + \cos B + \cos C + \cos D$ as a product of three terms.

How does this solve the problem?

(5) When you think you have managed to solve a problem completely, it is often a good idea to stand back and take a long hard look at what you have done. Read the question again *carefully*, check your working, think about what you actually *used* in achieving your solution, and so on. And don't be surprised if you discover a mistake: progress in mathematics often consists of taking two steps forward and one step backwards. Even when you don't find a mistake, you will often find that you *have* overlooked something! The original question refers to a *convex* quadrilateral; that is, a quadrilateral that does not have any 'dents'. Where does your solution make use of the fact that the quadrilateral $ABCD$ is *convex*?

[You might like to use the same ideas to find all quadrilaterals $ABCD$ with the property that $\sin A + \sin B + \sin C + \sin D = 0$—or to prove that none exist.]

---

4. A coin is biased so that the probability of obtaining a head is p, $0 < p < 1$. Two players, A and B, throw the coin in turn until one of the sequences HHH or HTH occurs. If the sequence HHH occurs first, then A wins. If HTH occurs first, then B wins. For what value of p is the game fair (that is, such that A and B have an equal chance of winning)?

---

(1) Probability questions are rarely 'easy'. But when dealing with a sequence of repeated tosses like this, one can often obtain a better picture of what is going on by drawing a simple tree-diagram like the one here. Circle those nodes where A wins and put a cross where B wins. Try to make sense of what you find.

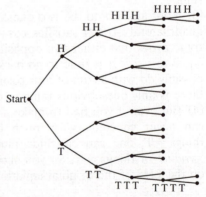

(2) Denote the probability that A wins the whole game by α and the probability that B wins by β. If the first toss is a tail, this helps neither A nor B. So the probability that A wins a game that has already started out 'T...' is still ____; similarly, the probability that B wins such a game is ____. Thus the probability that A wins a game and that the

first toss is T is $(1-p) \times$ ____ . Similarly, the probability that B wins a game and that the first toss is T is $(1-p) \times$ ____ .

(3) A can win in four ways:

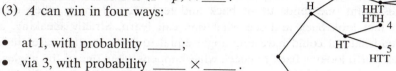

- at 1, with probability ____ ;
- via 3, with probability ____ $\times$ ____ .
- via 5, with probability ____ $\times$ ____ .
- via 6, with probability ____ $\times$ ____ .

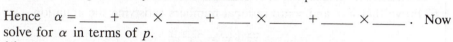

Hence $\alpha =$ ____ + ____ $\times$ ____ + ____ $\times$ ____ + ____ $\times$ ____ . Now solve for α in terms of p.

(4) Do the same for β. Finally, choose p such that $\alpha = \beta$.

[Give an example of two sequences of tosses for which neither A nor B wins. It should therefore come as a slight surprise that $\alpha + \beta =$ ____ . Can you prove this directly?]

---

5. The diagonals of a convex quadrilateral $ABCD$ intersect at O. The centroids of triangles AOD and BOC are P and Q; the orthocentres of triangles AOB and COD are R and S. Prove that PQ is perpendicular to RS.

---

(1) Every mathematics problem has to be understood before you can begin to solve it. A *convex* quadrilateral is one without dents (that is, with all angles $\leqslant 180°$). You are told that the quadrilateral $ABCD$ is convex. What does this tell you about the point O?

(2) Drawing a diagram proves absolutely nothing. But it is a very good way of making sure that you understand the problem. The very act of drawing can sometimes suggest ways of solving the problem. And an accurate drawing may reveal coincidences which give you other ideas. Draw an accurate diagram (for an arbitrary convex quadrilateral $ABCD$). Mark the triangles AOD, BOC, AOB, and COD. Check the meaning of *centroid* and *orthocentre* (in 'A little useful mathematics'). Mark P, Q, R, and S on your diagram. Make sure that PQ and RS look perpendicular: if they don't, your diagram must be inaccurate (or else the question is wrong).

(3) Perhaps the most obvious way of proving two lines are perpendicular is to use vectors and to show that some dot product is equal to zero. What would be a good point to choose as the origin? Suppose that the points A, B, C, D, R, and S have position vectors $\mathbf{a}$, $\mathbf{b}$, $\mathbf{c}$, $\mathbf{d}$, $\mathbf{r}$, and $\mathbf{s}$. Find the position vectors of P and Q in terms of these. What do you know about the lines OR and AB? What does this tell you about the vectors $\mathbf{r}$ and $\mathbf{a} - \mathbf{b}$? What do you know about $\mathbf{s}$ and $\mathbf{c} - \mathbf{d}$?

(4) Show that PQ is perpendicular to RS provided that $\mathbf{s} \cdot (\mathbf{a} - \mathbf{b}) = \mathbf{r} \cdot (\mathbf{d} - \mathbf{c})$.

Use the many pairs of perpendicular lines in the diagram (such as SC, BD and SD, AC) to show that $\mathbf{s} \cdot (\mathbf{a} - \mathbf{b}) = \mathbf{a} \cdot \mathbf{d} - \mathbf{b} \cdot \mathbf{c}$. Then do the same for $\mathbf{r} \cdot (\mathbf{d} - \mathbf{c})$, and hence solve the problem.

(5) At this point one tends to sit back and feel vaguely smug, when one should in fact look back and see what one can learn. Strictly speaking, a convex quadrilateral could have one angle equal to 180°. In such cases it is sometimes worth looking back to check what happens when one of the angles, say A, is *equal* to 180°? Where is the point O? Where is the point R? And where is the centroid P of triangle AOD?

---

6. Prove that if x and y are rational numbers satisfying the equation

$$x^5 + y^5 = 2x^2 y^2$$

then $1 - xy$ is the square of a rational number.

---

(1) Equations in rational numbers are a bit like equations in integers. But it is not always a good idea to go from an equation in *two* unknown *rationals* x and y to an equation in *four* unknown *integers* p, q, r, and s by writing $x = p/q$ and $y = r/s$. Instead, you should try to *simplify*.

(2) If x and y are rational numbers, then so is x/y. Although the given equation is not quite *homogeneous* (that is, having all terms of the same degree), it is tempting to divide both sides by y^5 to obtain

$$(x/y)^5 + 1 = 2(x/y)^2 / y.$$

Now put $x/y = t$. (If x and y are rational (and $y \neq 0$), then so is t.) Obtain an expression for y in terms of t. Obtain an expression for x in terms of t.

(3) Show that if t is any rational $\neq -1$, substituting these two expressions for y and x in the original equation gives a rational solution. How does this solve the original problem?

25th British Mathematical Olympiad, 1988

---

1. Find all integers a, b, and c for which

$$(x - a)(x - 10) + 1 = (x + b)(x + c) \qquad \text{for all } x.$$

---

(1) This is *not* an equation to be solved! It is meant to be an identity, with the two sides being equal *for all* x. Hence the two sides must be equal as *polynomials*, with the same coefficients on each side. The coefficient of x^2 on each side is obviously 1: so far, so good! How about the coefficient of x? To make those equal we need

$$-a - 10 = b + c.$$

And how about the constant terms on each side? To make those equal we must have

$$10a + 1 = bc.$$

That gives us *two* equations in *three* unknowns. If you eliminate a and take everything to one side you obtain *one* equation involving *two* unknowns, b and c, of the form $bc +$ ___ $b +$ ___ $c +$ ___ $= 0$. And then you seem to be stuck!

(2) Go back and read the question again: 'Find all *integers* a, b, and c...'. Aha! So a, b, and c are not any old unknowns; they have to be *integers*. The most important thing about integers is the way in which they factorize. Rewrite the equation obtained in (1) in the form

$$\overline{\hspace{6cm}} = 1.$$

Then factorize the LHS and find all solutions. (This method only works because we know that b and c are *integers*. If b and c were arbitrary *real* unknowns, you should automatically make the RHS $= zero$ before trying to factorize.)

(3) Now that you have managed to find a solution, it is time to reflect. Look back at the question. Can you see how your final equation '( )( ) $= 1$' could have been obtained directly from the identity given in the question? What value should you choose to substitute for x in the original identity?

---

2. Points P and Q lie on the sides AB and AC respectively of triangle ABC and are distinct from A. The lengths AP and AQ are denoted by x and y respectively, with the convention that $x > 0$ if P is on the same side of A as B, and $x < 0$ on the opposite side; and similarly for y. Show that PQ passes through the centroid of the triangle if and only if

$$3xy = bx + cy,$$

where $b = AC$ and $c = AB$.

---

(1) The whole question reeks of vectors. Which is the obvious point to choose for the origin O? Suppose that $\vec{AB} = \mathbf{c}$ has length c and $\vec{AC} = \mathbf{b}$ has length b. Express $\vec{AP}$ and $\vec{AQ}$ in terms of these.

(2) One basic property of vectors is that a point X lies on the line joining P and Q precisely when

$$\vec{OX} = \lambda\vec{OP} + (1 - \lambda)\vec{OQ}$$

for some real number λ: if $\lambda = 0$, then $X = Q$; if $\lambda = 1$, then $X = P$; if $\lambda = \frac{1}{2}$, then X is the midpoint of PQ. Let G be the centroid of triangle ABC. There is a simple expression for $\vec{AG}$ in terms of **b** and **c**. What is it? Use this to show that

$$\vec{AG} = \lambda\frac{x}{c}\mathbf{c} + (1-\lambda)\frac{y}{b}\mathbf{b}$$

(that is, that G lies on PQ) precisely when

$$3xy = bx + cy.$$

---

3. The lines OA, OB, and OC are mutually perpendicular. Express the area of triangle ABC in terms of the areas of triangles OBC, OCA, and OAB.

---

(1) The first move in any geometric question must be to draw a diagram. It is not easy to draw good 3-D diagrams. But in this case if we think of OA, OB, and OC as the (positive) x-, y-, and z-axes, it should be possible to produce something reasonable.

(2) What is 'special' about all three triangles OBC, OCA, and OAB? Your answer should suggest that the areas of these three triangles are best expressed in terms of the three lengths $OA = x$, $OB = y$, and $OC = z$. Use these lengths to find the areas of $\triangle OBC$, $\triangle OCA$, and $\triangle OAB$.

(3) How about the area of $\triangle ABC$? The simplest ways of finding the area of $\triangle ABC$ ($\frac{1}{2}$base $\times$ height, and $\frac{1}{2}ab\sin C$) require other information about $\triangle ABC$. We don't yet know the 'height', or any of the angles, but we should be able to find them now that we know the lengths of the three sides. Use $OA = x$ and $OB = y$ to find AB, and hence the length of the altitude OD from O to AB. Hence calculate the height CD of triangle ABC, and the area of triangle ABC. Find a formula giving the area of $\triangle ABC$ in terms of the areas of $\triangle OAB$, $\triangle OBC$, and $\triangle OCA$.

(4) This direct approach should succeed. An alternative is to use the formula (Heron's formula) which gives the area of triangle ABC purely in terms of the lengths of the three sides $AB = c$, $BC = a$, and $CA = b$. Let $2s = a + b + c$ be the perimeter of the triangle. Then

$$(\text{area } \triangle ABC)^2 = s(s-a)(s-b)(s-c).$$

Use the known expressions for a, b, and c (in terms of x, y, and z) to answer the original question. (The algebra may look horrible at first, but if you calculate intelligently the expression for (area $\triangle ABC)^2$ in terms of x, y, and z simplifies dramatically.)

(5) The formula 'area $= \frac{1}{2}ab\sin C$' may suggest a connection with vectors. The *dot* product (or *scalar* product) of two vectors **a** and **b** is a real number, and is very useful when trying to 'resolve' **a** (along **b** and perpendicular to **b**). This is *not* what we want here, since $\mathbf{a}\cdot\mathbf{b} = |\mathbf{a}|\,|\mathbf{b}|\cos\theta$ depends on the *cosine*

of the angle θ between the two vectors. However, there is another product of **a** and **b**—the *cross* product (or *vector* product), $\mathbf{a} \times \mathbf{b}$: this is a vector of length $|\mathbf{a}|\,|\mathbf{b}|\sin\theta$ (= the area of the parallelogram spanned by **a** and **b**) and with direction perpendicular to **a** and **b**. Hence

$$\text{area of triangle } ABC = \tfrac{1}{2}|\vec{AB} \times \vec{AC}|.$$

At first sight this may just look like a complicated way of saying 'area = $\frac{1}{2}bc\sin A$'. But it has one tremendous advantage: once you are used to it, calculating vector products is straightforward!

Let $OA = x\mathbf{i}$, $OB = y\mathbf{j}$, and $OC = z\mathbf{k}$, where **i**, **j**, and **k** are mutually perpendicular unit vectors (forming a right-handed system). Write down $\vec{AB}$ and $\vec{AC}$ in terms of **i**, **j**, and **k**. Use the fact that $\mathbf{i} \times \mathbf{j} = \mathbf{k} = -\mathbf{j} \times \mathbf{i}$, $\mathbf{j} \times \mathbf{k} = \mathbf{i} = -\mathbf{k} \times \mathbf{j}$, $\mathbf{k} \times \mathbf{i} = \mathbf{j} = -\mathbf{i} \times \mathbf{k}$, to work out $\vec{AB} \times \vec{AC}$ purely by algebra, without worrying about 'angles'. Hence answer the original question.

---

4. Consider the triange of numbers on the right. Each number is the sum of three numbers in the previous row: the number above it and the numbers immediately to the left and right of that number. If there is no number in one or more of these positions, 0 is used. Prove that, from the third row on, every row contains at least one even number.

$$
\begin{array}{ccccccccc}
 & & & & 1 & & & & \\
 & & & 1 & 1 & 1 & & & \\
 & & 1 & 2 & 3 & 2 & 1 & & \\
 & 1 & 3 & 6 & 7 & 6 & 3 & 1 & \\
 \cdot & \cdot & \cdot & \cdot & \cdot & \cdot & \cdot & \cdot & \cdot
\end{array}
$$

---

(1) There is only one obvious way to begin to get some feeling for this kind of simple-but-unfamiliar array: write out the first ten or so rows for yourself. Do you notice any patterns at all (perhaps familiar, perhaps not) in the array of numbers you produce?

(2) Each number in the next row only depends on the three numbers immediately above it. Hence the left–right symmetry in the first few rows is bound to continue. Thus, in each row, the first half is just the reverse of the last half. There are several other things that you should have noticed. The way in which the left-hand sloping edge is produced shows that it must consist entirely of 1's. The way in which the next sloping diagonal is generated shows that it must consist of the natural numbers $1, 2, 3, \ldots$ (Why?). You should recognize the numbers in the next sloping diagonal. Can you *prove* that the nth number in this sloping diagonal really is what you think it is?

(3) It follows from these initial observations that most rows have an even number in either the second or the third position. So, if we want, we can forget about those rows and concentrate on the rest. Suppose that we label

the rows by the sequence of natural numbers in the *second* sloping diagonal (the top row '1' is then the **0**th row; the next row '1 1 1' is the **1**st row; and so on.) For which values of n does the nth row have odd numbers '1, odd, odd, ...' in the first three positions?

(4) Look at the 5th and 9th rows. What do you notice about the numbers in the fourth position in each row?

(5) Of course, what you noticed for the 5th row and the 9th row may not be true for the 105th row and the 109th row. The original question is deliberately vague about *where* one should look in a row to find the even numbers. And the first ten rows tend to suggest that there is no simple general pattern. However, (4) may suggest that the original *vague* question could perhaps be resolved by proving something much more specific; namely, that the *fourth* number in the $(4n + 1)$th row is always even, for every $n \geqslant 1$. Let f_m denote the fourth number in the mth row:

- Show that $f_{m+1} = m + \frac{1}{2}m(m + 1) + f_m$.
- Use $f_m = \sum_{i=2}^{m}(f_i - f_{i-1})$ to find a formula for f_m. (Use the formula for $\sum_{i=1}^{m} i^2$.)
- Explain why this solves the original problem.

(6) Mathematicians, like politicians, succeed by applying the principle 'divide and rule'. The above approach starts by picking off the easy cases. Half the rows have an even number in the second position; and half the remaining rows have an even number in the third position. This allows one to concentrate on every fourth row—the $(4n + 1)$th rows, $n \geqslant 1$. This strategy can be very effective. But afterwards it is often worth looking back to see if there is a more uniform approach. If you really want to gain insight into the way odd and even numbers occur in this curious number triangle, you will have to look at much more than the first few rows. Unfortunately, the whole process of adding three numbers to find the next number gets in the way, and soon leads to errors. This is particularly annoying, since you don't really care whether the middle number in the eighth row is 1107 or 1109 as long as you know that it is *odd*. This suggests simply writing 'O' for odd and 'E' for even, and adding in the obvious way ($O + O = E$, $O + E = O$, and $E + E = E$). Take a large sheet of paper and write out the first 20 rows of the number triangle, entering 'O' for odd and 'E' for even entries. You already know that the two outside diagonals consist entirely of O's (namely 1's). What other striking patterns of O's and

$$O$$
$$O \quad O \quad O$$
$$O \quad E \quad O \quad E \quad O$$
$$O \quad O \quad E \quad O \quad E \quad O \quad O$$
$$\cdot \quad \cdot \quad \cdot \quad \cdot \quad \cdot \quad \cdot \quad \cdot \quad \cdot \quad \cdot$$

E's do you notice? Which of these patterns can you *prove* goes on for ever, as you suspect?

(7) You may have noticed one or two places where odd numbers seem to occur. It seems much harder to pin down anything very useful about what you really want to know—namely about where the *even* numbers occur. Still, you should at least have noticed, and been able to prove, that there is one vertical column which consists entirely of odd numbers. It is not clear how that might help, but perhaps you should bear it in mind.

(8) Whenever you want to prove something (such as 'every row after the first two contains at least one even number') and have no idea how to begin, there is one truly marvellous way of getting started, which has the double advantage of making the whole problem much clearer and of giving you additional information apparently for nothing. (Does this seem too good to be true? Well, there is a hitch; but the advantages are so impressive that you probably won't even notice.) You want to show that, except for the 0th and 1st rows, every row contains at least one even number. All that you know for sure is that the first, the last, and the middle number in each row is odd. Apart from that, you are pretty well stuck! So what do you do? Just relax! Stop trying to prove *directly* that every row must contain at least one even number. Instead, start thinking about what it would mean if the thing you want to prove true were actually *false*: that is, if *some* row were to contain *no even numbers at all*. If you can show that this is only possible for the first two rows, then you will be home and dry.

The advantage of this switch is that instead of trying to think about some arbitrary *n*th row, you can concentrate on one particular row about which you know something very specific; namely, that it consists entirely of odd numbers. That is, not only does the number triangle begin as you know it begins; but somewhere further down the triangle there would be a row consisting *entirely of odd numbers*. At this stage you do not know how far down the triangle this 'odd' row is, so you don't know how long the row is. But just suppose that there is a row *somewhere*. What can you say about the previous row? And what can you say about the row before that? How does this solve the original problem?

---

5. None of the angles of a triangle *ABC* exceeds 90°. Prove that

$$\sin A + \sin B + \sin C > 2.$$

---

(1) What an amazing inequality! Where on earth could it have come from? And how good is it? What is the value of the LHS for an equilateral triangle? What about an isosceles right-angled triangle? What about an isosceles triangle with a very small apex angle? What about your favourite acute-angled triangle? Can you find an *obtuse*-angled triangle for which the inequality is false?

(2) Well, how are you going to prove it? The LHS mentions all three angles *A*, *B*, and *C*; but angle *C* clearly depends on *A* and *B*. Thus it may seem

natural to substitute for C in terms of A and B, and to use the basic trig formulae to rewrite the LHS. Rewrite $\sin A + \sin B$ as a product, take out a common factor; then use the formula for $\cos X + \cos Y$ to express the LHS as a product; finally, use $\sin\frac{1}{2}(A+B) = \cos(C/2)$ to obtain the original LHS in the form $4\cos(A/2)\cos(B/2)\cos(C/2)$. Since the angles $A/2$, $B/2$, and $C/2$ are all $\leqslant 45°$, this product is clearly $> \sqrt{2}$. This shows that the inequality to be proved depends on the fact that, within the range $(0, 45°]$, we must somehow play off the exact value of $C/2$ against the values of $A/2$ and $B/2$.

(3) Go back to the less symmetrical expression for the LHS, namely $2\cos(C/2)(\cos\frac{1}{2}(A+B) + \cos\frac{1}{2}(A-B))$. For each given value of C, the value of the first factor '$\cos(C/2)$' and the value of $A+B$ are both fixed; hence we need to find a lower bound for $\cos\frac{1}{2}(A-B)$; that is, an upper bound for $\frac{1}{2}(A-B)$. Therefore, if C is fixed, $A+B$ ($= 180° - C$) is fixed, so to find a lower bound for $\cos\frac{1}{2}(A-B)$ we must take A as large as possible and B as small as possible; that is, $A =$ ___ , $B =$ ___ . To obtain a lower bound for the original LHS, we must therefore choose C ($< 90°$) so that $2\cos(C/2)(\sin(C/2) + \cos(C/2)) = \sin C + \cos C + 1$ is as small as possible. (For this, all you need is that, since $0 < C < 90°$, both $\sin C$ and $\cos C$ lie strictly between 0 and 1, so $\sin C + \cos C > \sin^2 C + \cos^2 C = 1$.)

[Alternatively, working in terms of radians rather than degrees, you might like to prove that when $0 < \theta \leqslant \pi/2$ we have $\sin\theta \geqslant 2\theta/\pi$ (with equality only for $\theta = \pi/2$). Hence $\sin A + \sin B + \sin C > (2/\pi)(A+B+C)$!]

---

6. The sequence $\{a_n\}$ of integers is defined by $a_1 = 2$, $a_2 = 7$, and

(*) $-\dfrac{1}{2} < a_{n+1} - \dfrac{a_n^2}{a_{n-1}} \leqslant \dfrac{1}{2}$ for $n \geqslant 2$.

Prove that a_n is odd for all $n > 1$.

---

(1) How curious! Given the inequalities for a_{n+1}, it is not at all clear how a_n behaves. The way in which a_{n+1} is defined seems completely inscrutable. Still, the only way of making it *less* inscrutable is to use it to work out the next few terms. The whole question takes it for granted that, once we know a_{n-1} and a_n, the inequalities should determine a_{n+1} *uniquely*. Use (*) with $n = 2$ to find a_3. Is it uniquely determined? Use (*) with $n = 3$ to find a_4. Is it uniquely determined?

(2) Find a_5, a_6, a_7, a_8, and a_9. Are they all odd?

(3) By now it should at least be clear that the inequalities defining a_{n+1} are not as complicated as they seemed. Once we know that a_{n-1} and a_n are integers, it follows that a_n^2/a_{n-1} is a rational number; a_{n+1} is then just the *nearest integer* to this rational number. But why on earth should this 'nearest

integer' always be *odd*? And why is the fraction a_n^2/a_{n-1} always so *close* to the nearest integer? Go back and work out the actual differences $a_{n+1} - a_n^2/a_{n-1}$ when $n = 2, 3, 4, 5, 6, 7$.

(4) At this stage you should 'smell a rat', even if you don't know how to catch it. Examine successive ratios a_{n+1}/a_n, $1 \leqslant n \leqslant 8$. Do you notice anything interesting about this sequence of ratios? Can you rearrange the inequality (∗) so that it involves successive terms of the sequence $\{a_{n+1}/a_n\}$?

(5) This more symmetrical version of the defining inequality may turn out to have certain advantages. But it still *conceals* rather than *reveals* why it might be true that 'a_n is odd for all $n > 1$'. There may be mathematical techniques that could help here, but if one doesn't know them, then one has to try something more straightforward. Could there be some simpler rule which defines our sequence $\{a_n\}$, perhaps like the rule for generating Fibonacci numbers? Look carefully at the ratios of successive terms

$$\frac{a_2}{a_1}, \frac{a_3}{a_2}, \frac{a_4}{a_3}, \frac{a_5}{a_4}, \ldots .$$

The fractional part of each ratio may be a little hard to understand; but there is something remarkably constant. This suggests that each a_{n+1} consists of some fixed multiple $k \cdot a_n$ of the previous term a_n, plus an extra bit. What looks like the most promising value of k? Can you make sense of the 'extra bit' that is needed each time? If you write out the sequence $\{a_{n+1} - ka_n\}$ of 'extra bits' required, it should not be long before you think you know what is going on.

(6) You should now have one official defining rule (∗) for the sequence $\{a_n\}$, and another much simpler rule *which seems to define the same sequence*. However, all of this is wild (one hopes, inspired) guesswork; you don't *know* that the two sequences are the same. They certainly start in the same way; but the defining rules look so different that this could easily be an accident. To keep the two sequences apart in your mind, give the new, simpler sequence a different *name* $\{b_n\}$: that is, $\{b_n\}$ is the sequence in which $b_1 = 2$, $b_2 = 7$, and b_{n+1} is given by the simple recurrence $b_{n+1} = 3b_n + 2b_{n-1}$. You have to prove that this is just an alternative description of the original sequence. But there is no point doing this unless you have some idea how this will help you to solve the original problem.

(7) Explain why b_n has to be odd for all $n > 1$. Now take a deep breath and try to show that $b_n = a_n$ for all n. Clearly, $\{b_n\}$ is a sequence of *integers*. Suppose that you know that $b_i = a_i$ for all $i \leqslant n$. Show that

$$-\frac{1}{2b_n} < \frac{b_{n+1}}{b_n} - \frac{b_n}{b_{n-1}} \leqslant \frac{1}{2b_n}.$$

Hence solve the original problem.

24th British Mathematical Olympiad, 1987B

---

1. Find all real solutions x of the equation

$$\sqrt{(x + 1\,972\,098 - 1986\sqrt{(x + 986\,049)})}$$
$$+ \sqrt{(x + 1\,974\,085 - 1988\sqrt{(x + 986\,049)})} = 1.$$

---

(1) This is Question 1 on the paper, so it ought to be reasonably easy. You may be tempted to try to get rid of the root signs by squaring both sides. If so, just stand back and think a little before you begin. How many times are you going to have to square things to get rid of *all* the square root signs?

(2) Suppose that $\sqrt{(a - c\sqrt{b})} + \sqrt{(d - e\sqrt{b})} = 1$. Square both sides. Then isolate the 'large' root on one side and square both sides again. Finally, equate coefficients on both sides; that is, equate the constant terms on each side, and equate the coefficients of $\sqrt{b}$ on each side. It looks as though this method may work. (There are technical problems in 'equating coefficients'. There is also the extra complication that each time we square both sides we introduce 'false' solutions.)

(3) Is there not some simpler way? Is it perhaps possible that the two long expressions under the square roots in the original question can actually be written as perfect squares? If so the outside square roots will just melt away.

(a) Try to see whether we might in fact be able to write

$$x + 1\,972\,098 - 1986\sqrt{(x + 986\,049)} = (\underline{\hspace{6cm}})^2.$$

Can you work out what has to go in the bracket on the RHS?

(b) Can you do the same for the other expression

$$x + 1\,974\,085 - 1988\sqrt{(x + 986\,049)} = (\underline{\hspace{6cm}})^2?$$

(4) This means that the LHS of the original equation simplifies considerably. However, one must be careful about simply taking square roots, since the expressions would go in the brackets on the RHS of the identities in (3)(a) and (3)(b) could easily be *negative*, and '$\sqrt{}$' indicates the *non*-negative root. The only way to avoid a mistake here is to use modulus signs:

$$\sqrt{(\dots)^2} = |(\dots)|.$$

The equation to be solved now takes the form

$$|y - \alpha| + |y - \beta| = 1.$$

What is y? What are α and β?

(5) Solving a modulus equation—even a simple looking one—requires considerable care. The key is to realize that

'$|y - \alpha|$' means 'the distance from y to α'.

Show that $|y - \alpha| + |y - (\alpha + 1)| = 1$ precisely when y lies between α and $\alpha + 1$. Use this to solve the original equation.

[Alternatively, you might prefer to change variable by putting $x + 986\,049 = u$ right at the start.]

---

2. Find all real-valued functions f which are defined on the set D of natural numbers $x \geqslant 10$, and which satisfy the functional equation

$$f(x + y) = f(x) \cdot f(y),$$

for all $x, y \in D$.

---

(1) Which *familiar* functions f have the property that $f(x+y) = f(x) \cdot f(y)$? Your first guess has to be that the unknown functions f in the question may be just these familiar functions in disguise: after all, the functional equation is just a fancy way of writing one of the familiar index laws for powers. But why does the question restrict to integers $x \geqslant 10$? Is this merely a smoke-screen? Or could you have missed something? Perhaps things will become clearer once you get going.

(2) Let f be an arbitrary *unknown* function with the property that $f(x+y) = f(x) \cdot f(y)$ for all integers $x \geqslant 10$. What can you say about $f(2x)$? About $f(3x)$? About $f(nx)$?

(3) Suppose that $f(10) = a$. What can you say about $f(20)$? About $f(30)$? About $f(110)$? About $f(x \cdot 10)$?

(4) What does $f(3x) = f(\frac{3}{2}x + \frac{3}{2}x)$ tell you about $f(\frac{3}{2}x)$? About $f(\frac{6}{5}x)$? About $f(\frac{11}{10}x)$? About $f(\frac{5}{2}x)$? Suppose that $f(10) = a$. What can you say about $f(15)$? About $f(12)$? About $f(11)$? About $f(25)$? Does the value $f(10) = a$ determine the value of $f(x)$ for all natural numbers $x \geqslant 10$? Now complete the solution. (Let $f(x) = b$. Then $f(2x) = \underline{\quad}$ and $f(10x) = \underline{\quad}$. But $f(x \cdot 10) = \underline{\quad}$ from (3). Hence $b = \underline{\quad}$.)

---

3. Find a pair of integers r, s such that $0 < s < 200$ and
$$\frac{45}{61} > r/s > \frac{59}{80}.$$

Prove that there is exactly one such pair r, s.

---

(1) At first glance, one feels that it should be easy to find a rational number r/s between $\frac{45}{61}$ and $\frac{59}{80}$. Check that $\frac{60}{80} = \frac{3}{4} = \frac{45}{60} > \frac{45}{61} > \frac{59}{80}$. It follows that the

required number r/s should lie between $\frac{59}{80}$ and $\frac{60}{80} = \frac{3}{4}$. Since you want $s < 200$, it is natural to try $59\frac{1}{2}/80 = 119/160$; this certainly lies between $\frac{59}{80}$ and $\frac{60}{80}$, and has a denominator $s < 200$. Does it lie between $\frac{59}{80}$ and $\frac{45}{61}$, or between $\frac{45}{61}$ and $\frac{45}{60}$?

(2) Notwithstanding this initial disappointment, it may be worth trying to find a rational number r/s of the required type by intelligent trial and error. In a problem like this it is important to have a go, and to reflect on what you find. You may be lucky and succeed relatively quickly. Even if the problem turns out to be harder than you expected, you should still have gained some valuable insight into *why* the problem is perhaps harder than it looks. In this case the difficulty lies in the fact that the upper and lower bounds $\frac{45}{61}$ and $\frac{59}{80}$ are surprisingly close. Work out $\frac{45}{61} - \frac{59}{80}$ as a fraction.

(3) The chances that a fraction with denominator < 200 should lie in this very small gap are pretty slim. However, there is a standard fact which can often be quite useful. Suppose that $a/b < c/d$, where $b, d > 0$. Where does $(a+c)/(b+d)$ lie relative to a/b and c/d? (Does this depend on the values of a, b, c, and d? Or can you prove that $(a+c)/(b+d)$ is 'always $< a/b$'? Or is it 'always $> c/d$'? Or is it 'always strictly between a/b and c/d'?)

(4) Step (3) should certainly allow you to find a pair of integers r, s with the required property. The sting in the tail is that you now have to show that this pair is the *only possible* pair with $0 < s < 200$. This depends on something you already noticed in step (2); namely, that $45 \times 80 - 59 \times 61 = 1$. Suppose that $a/b < c/d$ with $cb - ad = 1$. Then step (3) guarantees that $(a+c)/(b+d)$ lies between a/b and c/d. Suppose that x/y also lies between a/b and c/d. Show that the extra condition $cb - ad = 1$ implies that $y \geqslant b + d$. ($x/y < c/d$ implies that $cy - xd \geqslant 1$, and $x/y > a/b$ implies $-xd \leqslant -(ady + d)/b$; hence $1 \leqslant cy - xd \leqslant (y - d)/b$.)

(5) Now use step (4) three times (for $\frac{59}{80} < \frac{45}{61}$, for $\frac{59}{80} < \frac{104}{141}$, and for $\frac{104}{141} < \frac{45}{61}$) to complete your solution of the original problem.

---

4. The triangle ABC has orthocentre H. The feet of the perpendiculars from H to the internal and external bisectors of angle BAC (which is not a right angle) are P and Q. Prove that PQ passes through the midpoint of BC.

---

(1) While you may know a little about the circumcentre O (where the perpendicular bisectors of the three sides meet), the incentre I (where the angle bisectors meet), and the centroid G (where the medians meet), the *orthocentre* H—where the perpendiculars AA', BB', and CC' from each vertex to the opposite side, the 'altitudes', meet—is somehow less memorable. Still, you must not let that frighten you. Draw a diagram!

(2) The diagram tends to be a little messy and gives little clue as to *why* PQ should pass through the midpoint of BC. The most striking thing is all those

right angles produced by the altitudes AA', BB', and CC' and the rectangle $HPAQ$. The right angles in the figure described in the question seem to be scattered all over the place, with no obvious connections. You need an idea to link them. What simple geometrical shape produces lots and lots of related right angles?

(3) The diagram is full of hidden ∗i∗∗∗e∗. The circle with diameter PQ establishes a connection between six different right angles in your diagram. What are they? Since $\angle BB'C$ is a right angle, the point B' lies on the semicircle with diameter BC (and the important midpoint of BC is the ∗e∗∗∗e of this circle). Which other right angle in the diagram lies on this same semicircle? The supplementary right angle $\angle BB'A$ at B' lies on the two other hidden semicircles. What are their diameters?

(4) You have to prove something about the line PQ, so it seems sensible to begin, as in (3), by looking for a circle which involves the right angles at both P and Q. (There is only one real choice.) Which six labelled points in your diagram lie on the circumference of this circle? (Four of these points are obvious; to find the last two, observe that the circle has diameters PQ and ____.) Where is the centre of this circle? What can you say about $\angle B'AP$ and $\angle B'C'P$ (subtended by the same ∗∗o∗∗ $B'P$)? What can you say about $\angle C'AP$ and $\angle C'B'P$? What can you conclude about triangle $PB'C'$ (given that AP bisects $\angle BAC$)?

(5) You have to prove that PQ passes through the centre of the circle with diameter BC. In which two points does this circle meet the circle with diameter PQ? Use the fact that $PB' = PC'$ to prove that PQ is ∗e∗∗e∗∗i∗u∗a∗ to $B'C'$. Hence complete the solution.

---

5. For any two integers m and n with $0 \leqslant m \leqslant n$, numbers $d(n, m)$ are defined by

$$d(n, 0) = d(n, n) = 1 \qquad \text{for all } n \geqslant 0,$$

and

$$md(n, m) = md(n - 1, m) + (2n - m)d(n - 1, m - 1) \qquad \text{for } 0 < m < n.$$

Prove that all of the $d(n, m)$ are integers.

---

(1) There are times in mathematics when one has to take a deep breath and get started, in the hope that things will become clearer once you get stuck in. They certainly won't get any clearer if you avoid getting stuck in! Your only hope is to read the question carefully in the hope of recognizing something vaguely familiar in this welter of symbols. (John von Neumann, one of the greatest mathematicians of the twentieth century, expressed this very well when a student complained that he didn't understand something. 'Young

man', said von Neumann, 'in mathematics you don't have to understand things. You just get used to them.') The first condition, '$d(n,0) = d(n,n) = 1$ for all $n \geqslant 0$', should immediately remind you of binomial coefficients ($\binom{n}{0} = \binom{n}{n} = 1$ for all $n \geqslant 0$). The second condition then looks like a complicated variation on the rule for generating Pascal's triangle. Like binomial coefficients, the numbers $d(n,m)$, $n \geqslant m \geqslant 0$, can be arranged in a triangular array. For convenience, imagine this array in the first quadrant with the number $d(n,m)$ ($n \geqslant m \geqslant 0$) at the point with co-ordinates (n,m). Enter the numbers $d(n,0) = 1$ and $d(n,n) = 1$ (all $n \geqslant 0$) in your array.

(2) Triangular arrays of numbers bordered by 1's are not uncommon in mathematics. As with Pascal's triangle, there is usually some rule which allows one to calculate an unknown entry in terms of 'earlier' entries in the triangle. The rule in the question is just like the rule for Pascal's triangle, except that we have to add suitable *multiples* of two entries in the previous column:

* Calculate $d(2,1)$, $d(3,1)$, $d(4,1)$, and $d(5,1)$; enter the values in your array.
* Calculate $d(3,2)$, $d(4,2)$, and $d(5,2)$; enter the values in your array.
* Calculate $d(4,3)$ and $d(5,3)$; enter the values in your array.

(3) The numbers $d(n,m)$ with $m = 1$ ($n \geqslant 1$) should be thoroughly familiar. You are told that $d(1,1) = 1$. Suppose that $m = 1$ and $n \geqslant 2$; then the first term on the RHS of the recurrence $d(n,1) = \ldots + \ldots$ is '$1 \cdot d(n-1,1)$', while the second is just '$(2n-1) \cdot 1$'. It follows that $d(n,1)$ is obtained by adding the first n odd numbers (for which there is a very simple formula). The numbers $d(n,2)$ ($n \geqslant 2$) are harder to work out, but the answers should still make you suspicious. Calculating $d(4,3)$ and $d(5,3)$ is a little messy; but the answers should make you even more suspicious.

(4) So far, the numbers $d(n,m)$ all seem to be ∗∗ua∗e∗. But what are they squares of? It should not take you too long to guess the answer to this question: after all, you only know one other number triangle of this kind to compare it with. This should leave you in the position of being almost certain what $d(n,m)$ is in fact equal to. If you could *prove* that $d(n,m)$ is what you think it is then this would certainly show 'that all the $d(n,m)$ are integers'. Use induction (on the sum $n+m$) to prove that '$d(n,m) = \binom{n}{m}^2$' whenever $n \geqslant m \geqslant 0$.

(i) If $n+m = 0$, then $n = m = 0$ and $d(n,m) = d(0,0) = 1$, so $d(n,m) = \binom{n}{m}^2$.

(ii) Now suppose that $d(n, m) = \left(\dfrac{n}{m}\right)^2$ whenever $n + m < K$, and try to prove that when $n + m = K$ the same result holds (use the recurrence in the question, and the induction hypothesis that $d(n - 1, m) = \left(\dfrac{n-1}{m}\right)^2$ and $d(n - 1, m - 1) = \left(\dfrac{n-1}{m-1}\right)^2$).

[An alternative to this 'direct proof by induction' is to observe that the number triangle $d(n, m)$, $n \geqslant m \geqslant 0$, certainly agrees with $\left(\dfrac{n}{m}\right)^2$ on the two sloping edges of 1's. So if you simply check that $\delta(n, m) = \left(\dfrac{n}{m}\right)^2$ satisfies the same recurrence

$$m\delta(n, m) = m\delta(n - 1, m) + (2n - m)\delta(n - 1, m - 1) \qquad \text{for } 0 < m < n,$$

then the two number triangles $d(n, m)$ and $\delta(n, m)$ must be identical. All of this should leave you with the question of whether there is some way of proving 'that all the $d(n, m)$ are integers' without calculating the exact value of every $d(n, m)$ first.]

---

6. Show that, if x and y are real numbers such that $7x^2 + 3xy + 3y^2 = 1$, then the least positive value of $(x^2 + y^2)/y$ is $\frac{1}{2}$.

---

(1) Don't panic! The last question on each paper is meant to sort out the sheep from the goats. But that doesn't mean that it is necessarily terribly hard. There may well be fancy ways of tackling questions of this kind, but if you stick to simple-minded methods then there is really only one way to begin. Suppose that you wanted to show that, for all values of t, $t^2 - 4t \geqslant -4$. Any inequality like this can be rearranged into the form 'new LHS $\geqslant 0$'. And the new LHS is then a perfect square. What is it the square of?

(2) In easy exercises (such as proving that $t^2 - 4t + 4 \geqslant 0$ for all t) the LHS factorizes very simply as a single perfect square. Life is not always so kind. However, if you can prove that '$t^2 - 4t \geqslant -4$ for all t', then you can certainly prove that '$t^2 - 4t \geqslant -5$ for all t'. The second inequality may at first seem *harder*, but to show that $t^2 - 4t + 5 \geqslant 0$ for all t, it is enough to write the LHS as the *sum* of *two* squares: namely $(t - 2)^2 + 1^2$.

(3) Now use this simple idea to tackle the original problem. To find the least positive value of $(x^2 + y^2)/y$ you only have to worry about certain values of y. Which ones? Rearrange the required inequality '$(x^2 + y^2)/y \geqslant \frac{1}{2}$' into the form 'new LHS $\geqslant 0$'.

(4) Somehow you have to use the supplementary relation $7x^2 + 3xy + 3y^2 = 1$ to simplify the new LHS. (That is, you have to prove that for points above the x-axis on the curve with equation $7x^2 + 3xy + 3y^2 = 1$, the value of $(x^2 + y^2)/y - \frac{1}{2}$ is never negative.)

Unfortunately, the new LHS has an awkward y in the denominator; and there is no easy way to use an expression such as $7x^2 + 3xy + 3y^2$ *of degree 2* to simplify a single y *of degree 1*. The trouble is that the expression $(x^2 + y^2)/y$ has a numerator of degree 2 and a denominator of degree 1, so the whole expression $(x^2 + y^2)/y = x^2/y + y$ behaves in some ways as if it had degree $2 - 1 = 1$. This makes it difficult to write $(x^2 + y^2)/y$ in terms of an expression such as $7x^2 + 3xy + 3y^2$ of degree 2.

The simplest potential escape from this dilemma is to square our *original* inequality: for $y > 0$, $(x^2 + y^2)/y \geq \frac{1}{2}$ is equivalent to $[(x^2 + y^2)/y]^2 \geq \frac{1}{4}$. This approach may not work, but it is definitely the first thing to try. Rearrange the inequality '$[(x^2 + y^2)/y]^2 \geq \frac{1}{4}$' into the form 'new LHS ≥ 0', getting rid of denominators. You must now use the supplementary relation $1 = 7x^2 + 3xy + 3y^2$ to simplify the LHS, with a view to proving that the LHS is ≥ 0.

(5) There are many possible ways of using the supplementary relation $1 = 7x^2 + 3xy + 3y^2$ to rewrite the numerator. Most of these produce an expression which does not simplify in any obvious way. One natural substitution is to replace the coefficient '1' of y^2 in the expression '$4(x^2 + y^2)^2 - 1 \cdot y^2$' by '$7x^2 + 3xy + 3y^2$'. This may look as though it could only make things worse; but try it and see what happens. Then (remembering the point made in (2) above) try to finish off the solution.

(6) To complete the solution successfully you have to notice slightly more than was explicitly stated in (2). In proving $4x^4 + x^2y^2 + y^4 - 3xy^3 \geq 0$, you would like to write the LHS as a sum of squares. However, since there exist values $x = x_0$ and $y = y_0$ satisfying $7x^2 + 3xy + 3y^2 = 1$ which make the LHS $= 0$, the squares have to be chosen carefully: there is no slack. Find x_0 and y_0.

It follows that each of the squares that you use to rewrite the LHS must vanish when $x = x_0$ and $y = y_0$. In particular, since $y_0 = 2x_0$, this suggests that each square should involve $(2x - y)^2$. Expand $(4x^2 - y^2)^2 + 3y^2(2x - y)^2$. How does this solve the problem?

[Alternatively, let $x/y = k$. Then $y^2 = 1/(7k^2 + 3k + 3)$ and the inequality to be proved is just $4k^4 + k^2 - 3k + 1 \geq 0$. The solution can then be completed either by factorizing and completing the square, or by using calculus.]

23rd British Mathematical Olympiad, 1987A

---

1. (a) Find, with proof, all integer solutions of

$$a^3 + b^3 = 9.$$

(b) Find, with proof, all integer solutions of

$$35x^3 + 66x^2y + 42xy^2 + 9y^3 = 9.$$

---

(1) Part (a) looks reasonably easy; you can probably see two obvious solutions, $a = 2$, $b = 1$ and $a = 1$, $b = 2$. Moreover, it doesn't look as if there could be any others—although you will have to *prove* this later. How about part (b)? That 9 on the RHS looks suspiciously like the 9 in part (a). Could the LHS in part (b) possibly be simpler than it looks? Could it even be *exactly the same as the LHS in part (a)*, but heavily disguised: that is, of the form $a^3 + b^3 = 9$, with

(∗) $a = \alpha x + \beta y, \qquad b = \gamma x + \delta y?$

If so, you would have to have $\alpha^3 + \gamma^3 = 35$, $\beta^3 + \delta^3 = $ _____. There may be many solutions (since α, β, γ, and δ could be negative). However, it is tempting to try $\alpha = 2$, $\gamma = $ _____ (or $\alpha = 3$, $\gamma = $ _____) and $\beta = 1$, $\delta = $ _____ (or $\beta = 2$, $\delta = $ _____). One of these works. Which one is it? So once you know all integer solutions a, b to part (a), you can substitute these values of a, b in (∗) and find all integer solutions x, y to part (b).

(2) So perhaps the 'easy-looking' part (a) is, in some sense, the worst bit! Suppose that $a^3 + b^3 = 9$, with a and b both *positive*. Show that $0 < a^3$ and $b^3 < 9$, and that there are just two solutions.

(3) But what if b (say) is negative ($b = -\beta$) and a (say) is positive. Then

$$a^3 - \beta^3 = 9 \qquad \text{with } a, \beta > 0.$$

Can two positive cubes a^3 and β^3 differ by 9? Explain. Hence complete the solution to part (a).

---

2. In a triangle ABC, $\angle BAC = 100°$ and $AB = AC$. A point D is chosen on the side AC so that $\angle ABD = \angle CBD$. Prove that $AD + DB = BC$.

---

(1) Pictures are much easier to understand than all those words and letters! So draw a diagram and mark the information given on it.

(2) The most striking thing is that you seem to know so many angles. Indeed, there are so many angles that one is inclined to overlook crucial facts, such as $AB = AC$ and $\angle ABD = \angle CBD$. Still, when one knows lots of angles, one obvious strategy is to use trigonometry to find AD, DB, and BC, and then use trig identities $(\cos(90 - \alpha) = \sin \alpha, \ \sin 2\alpha = 2 \sin \alpha \cos \alpha, \ \cos \alpha - \cos \beta = \dots)$ to try to prove that $AD + DB = BC$. Trigonometry tends to be a little easier if you can work with right-angled triangles. Is there a natural way of introducing right-angled triangles, given that BD bisects the angle $\angle ABC$?

(3) An angle bisector is equidistant from the two legs of the angle that it bisects. Hence D is equidistant from BC and from BA (extended). Drop perpendiculars DE from D to BC, and DF from D to BA extended. Use the fact that $DE = DF$ to find x and y such that

$$AD = x \cdot BD, \qquad BC = y \cdot BD.$$

Then use trig identities to show that $AD + DB = BC$.

(4) You should have managed to make this first approach work, but it is a little messy. What is worse, it does not really explain *why* $AD + DB$ is equal to BC. When the relationship to be proved is as simple as $AD + DB = BC$, one feels that there should be a simple geometrical proof, although one can never be sure! (The trouble with simple proofs is that they are only simple *after* you have found them, and it can take an awful lot of searching before you hit on the right idea—the key that unlocks the whole puzzle.) The important things to notice here are that:

- the segments AD and DB are not in a line, so they are difficult to add together in a simple-minded way;
- neither of these segments lies along BC (so it looks as though you will definitely have to move something);
- the two 20° angles at B are *equal*.

Rotate triangle $BA(D)C$ through 20° to triangle $BA'(D')C'$ with A' lying on BD. Where does D' lie? What do you know about BD'? So what must you prove about $D'C$?

(5) By now your diagram may well be rather cluttered, but you should still be able to work out almost any angle you want. Join AA'. What can you say about the three points A, A', and D'? What can you say about triangle ADD'? What can you say about triangle $DD'C$? How does this help you solve the original problem?

---

3. Find, with proof, the value of the limit as $n \to \infty$ of the quotient

$$\sum_{r=0}^{n} \binom{2n}{2r} \cdot 2^r \bigg/ \sum_{r=0}^{n-1} \binom{2n}{2r+1} \cdot 2^r.$$

---

(1) This looks horrid! There's no way in which you can *think* about an expression like this until you can begin to see what it really *means*. Where on earth could it have come from? Forget about the exact form of the expression for the moment. Suppose that the numerator were $\sum_{r=0}^{n} \binom{n}{r} \cdot 2^r$. Would that be better? Of course it would: the binomial theorem says that

$$(1+x)^n = \sum_{r=0}^{n} \binom{n}{r} \cdot x^r;$$

$$\therefore \sum_{r=0}^{n} \binom{n}{r} \cdot 2^r = (1+2)^n = 3^n.$$

(2) This tells you the value of $\sum_{s=0}^{2n} \binom{2n}{s} \cdot 2^s$. However, the expression in the original problem has two awkward features:

- Only the *even* binomial coefficients $\binom{2n}{2r}$ occur in the numerator, with the *odd* ones occurring in the denominator.

- The binomial coefficient $\binom{2n}{2r}$ occurs with 2^r, not with 2^{2r} as you might expect.

The second of these is perhaps the easiest to understand. You expect $\binom{2n}{2r}$ to occur as the coefficient of x^{2r}. If $2^r = x^{2r}$, what must the value of x be? (3) This suggests that the whole problem has something to do with the binomial expansion of something like $(1 \pm \sqrt{2})^{2n}$. But you still have to explain why there are no $\sqrt{2}$s around, and why the *odd* terms $\binom{2n}{2r+1} 2^r$ are in the *denominator*. (These two puzzles are clearly related! After all, you would not expect to find any $\sqrt{2}$s with the *even* terms $\binom{2n}{2r} 2^r$!) Suppose that $\left(\sum_{r=0}^{n} \binom{2n}{2r} 2^r \middle/ \sum_{r=0}^{n-1} \binom{2n}{2r+1} 2^r \right) = t$. The value of t actually depends on n; but you are interested in the value as $n \to \infty$. Suppose that you cross-multiply and shift all the terms to the LHS. What value of t would allow you to simplify the LHS nicely using the Binomial Theorem? And what does $(1 - \sqrt{2})^{2n}$ tend to as $n \to \infty$? Now use this idea to solve the original problem.

---

4. Let $P(x)$ be any polynomial with integer coefficients such that

$$P(21) = 17, \qquad P(32) = -247, \qquad P(37) = 33.$$

Prove that if $P(N) = N + 51$, for some integer N, then $N = 26$.

---

(1) This whole question looks very strange. What do we know about polynomials? Not much—except for the *e*ai**e* theorem. Well, can we use the *e*ai**e* theorem? In its simplest form the theorem states that if $Q(x)$ is a polynomial and a is a real number, then

$$Q(a) = 0 \iff Q(x) = (x - a) \cdot R(x).$$

Most introductory texts leave it at that, but the usual proof actually proves a little more than this.

'If $Q(x) = a_n x^n + a_{n-1} x^{n-1} + \ldots + a_1 x + a_0$,

then $Q(a) = a_n a^n + a_{n-1} a^{n-1} + \ldots + a_1 a + a_0$.

$\therefore Q(x) - Q(a) = a_n(x^n - a^n) + a_{n-1}(x^{n-1} - a^{n-1}) + \ldots + a_1(x - a)$.

Now $x^i - a^i = (x - a)(x^{i-1} + x^{i-2}a + \ldots + a^{i-1})$ for each i.

$\therefore Q(x) - Q(a) = (x - a) \times$ (some polynomial $R(x)$),

where the coefficients of $R(x)$ are *integer* combinations of the coefficients of $Q(x)$ and powers of a. Hence *if a is an integer and $Q(x)$ has integer coefficients, then $R(x)$ has integer coefficients too*!'

(2) It may not be clear at first how to use the Remainder Theorem with the polynomial $P(x)$ itself. The Remainder Theorem applies to polynomials $Q(x)$ and numbers a for which $Q(a) = 0$, whereas Question 4 requires us to work with the condition '$P(N) = N + 51$ for some N'. There is an easy way out of this. You have to define a new polynomial $Q(x)$ (closely related to $P(x)$ so that $P(N) = N + 51 \Leftrightarrow Q(N) = 0$. Now work with this new polynomial $Q(x)$. Find $Q(21) =$ _____, $Q(32) =$ _____, and $Q(37) =$ _____.

(3) The fact that $Q(21) = Q(37) \neq 0$ suggests that we may have chosen to work with the wrong polynomial $Q(x)$. If we let $Q^*(x) = Q(x) + 55$, then $Q^*(21) = Q^*(37) = 0$; so by the Remainder Theorem $x - 21$ and $x - 37$ are both factors of $Q^*(x)$ and

$$Q^*(x) = (x - 21)(x - 37) \cdot R(x),$$

where $R(x)$ still has integer coefficients. Moreover,

$$P(N) = N + 51 \Leftrightarrow Q(N) = 0 \Leftrightarrow Q^*(N) = \underline{\quad}.$$

Can you see how this helps you to solve the original problem? (When you have succeeded, you might like to reflect on what role the information '$P(32) = -247$' has played in your solution.)

---

5. A line parallel to the side BC of an acute angled triangle ABC cuts the side AB at F and the side AC at E. Prove that the circles on BE and CF as diameters intersect on the altitude of the triangle drawn from A perpendicular to BC.

---

(1) Draw a diagram and mark all the information given.
(2) In mathematics one should always be on the lookout for tell-tale clues, which look as though they are going to be important in the eventual solution of a problem. In this particular problem the only obvious hint of this kind is the fact that altitudes and semicircles both involve *i*** a***e*. Of course this may turn out to be irrelevant, but you should bear it in mind when looking for a way into the problem. (School mathematics tends to overemphasize the idea that every problem includes a more or less obvious clue as to how to solve it. Real mathematical problems are much more interesting than this, and the important ideas generally emerge gradually as you begin to get to grips with the problem.) Suppose that the two circles on BE and CF as diameters intersect at P and Q. Then P is one of the two points for which $\angle BPE = \angle CPF = 90°$. We have to show that these conditions *force* P to lie on the altitude from A to BC. In other words, we have to show that AP is perpendicular to BC. There may be a clever way of doing this, but it looks as if we have not yet made enough use of the other information in the question

(for example, *FE* parallel to *BC*). In geometry questions, once one understands roughly what the diagram should look like, but still cannot quite see how to proceed, it is often a good idea to draw a more accurate diagram. Draw an accurate diagram with *BC* and *FE* horizontal and the altitude from *A* to *BC* vertical.

(3) Solving a mathematical problem is a process of pulling yourself up by your own bootstraps—one small step at a time. You should not expect to see straight through a good problem in one move. Instead, you should be permanently on the lookout for small steps that seem to lead in the right direction. Whatever your accurate diagram may suggest, it isn't at all obvious why *P* and *Q* should lie *exactly* on the altitude from *A* to *BC*. However, if you stare at your diagram (and think!) you should be able to see something which is definitely interesting. The common chord *PQ* of the two circles has to be 'vertical', that is, *parallel to* the altitude from *A* to *BC*. Can you see why? (The common chord of two intersecting circles is always perpendicular to the line joining their centres. So all you have to prove is that the line joining the midpoints of *BE* and *CF* has to be parallel to *BC*.)

(4) This certainly *feels* like progress. But the final step is still elusive. You now know that the altitude through *A* and the common chord *PQ* are parallel. It remains to be proved that they are not only parallel, but identical! In other words, you have to show that the point *A* has to lie *on* the common chord *PQ*. In solving any mathematical problem there comes a point at which one simply has to *know* something, or work it out from scratch. You want to know that *A* lies on the common chord *PQ* of our two circles. It is therefore hard to see how you can avoid the question

> 'What condition guarantees that a point *A* lies on the common chord of two intersecting circles?'

If you don't already know the answer, the next hint should help you work it out for yourself.

(5) Suppose that the point *A* lies outside a given circle, and that the line *AXY* cuts the circle in *X* and *Y*, while *AT* is tangent to the circle at *T*. Show that $\angle ATX = \angle TYX$. Conclude that triangles *ATX* and *AYT* are similar. Hence show that $AX \cdot AY = AT^2$. Deduce that a point *A* outside *two* intersecting circles lies on their common chord precisely when the lengths of the tangents from *A* to the two circles are equal. Try to see how this might help to solve the original problem. (If you succeed, then fine. If not, read on.)

(6) Your diagram does not contain a 'tangent *AT*' from *A* to the circle with diameter *BE*. Moreover, drawing one in would seem to make things worse. Thus the condition for *A* to lie on a common chord has to be used intelligently. Perhaps the easiest way is to forget about 'tangents', and to use the full force of the result in (5). The most obvious line from *A* which cuts the circle with diameter *BE* is *AB*; if this cuts the circle again at *X*, then

$AX \cdot AB = AT^2$ (even though we do not know where T is). Similarly, the most obvious line from A which cuts the circle with diameter CF is AC; if this cuts the circle again at Y, then you need only show that

$$AX \cdot AB = AY \cdot AC.$$

(For this, first observe that triangles ABC and AFE are *i*i*a* (Why?), so $AB/AC = AF/AE$. Then combine this with the fact that $AF/AY = AE/AX$ —from the similar right-angled triangles AFY and AEX.)

---

6. If x, y, and z are positive real numbers, find, with proof, the maximum value of the expression

$$\frac{xyz}{(1+x)(x+y)(y+z)(z+16)}.$$

---

(1) Each of the three factors x, y, and z in the numerator occurs in *two* separate factors in the denominator. So if you try to make the whole expression larger by increasing the numerator, then as x, for example, tends to infinity, the whole expression actually tends to zero! If, on the other hand, you try to make the whole expression larger by making the denominator small, then it is not enough to make just *one* of the variables small since, for example, x and y occur together in the same bracket; you have to make both x and y (or both y and z) small at the same time. But then the numerator contains the product $x \cdot y$ (or $y \cdot z$), so the whole expression actually becomes very *small!* Thus it looks as though the values of x, y, and z which give the expression its maximum value will be neither very large nor very small. So try some values which are neither very large nor very small. When $x = y = z = 1$, the expression has the value $1/(2 \cdot 2 \cdot 2 \cdot 17)$. What values does the expression have when $x = y = z = 2$? Is this larger or smaller than $1/(2 \cdot 2 \cdot 2 \cdot 17)$? What if $x = 4$, $y = 8$, and $z = 16$? Can you produce an even larger value?

(2) The previous step may give you a feeling for the approximate size of the maximum value (rather small!). But it proves nothing. If you have never seen a problem like this before, it may not be at all clear what to do next. So you really have no alternative but to try something simple-minded and see what happens.

One awkward feature of the given expression is that it doesn't appear to simplify very easily. But suppose that you turn the whole expression upside down! You will then have to look, not for the maximum value of the original expression, but for the *i*i*u* value of

$$\frac{(1+x)(x+y)(y+z)(z+16)}{xyz}.$$

At this point you must suppress any misguided urge to multiply out and collect up terms. Factorized form is generally far more useful, and multiplying out is often the worst possible thing to do with such an expression. However, you could get rid of the denominator by incorporating the 'x' in the second bracket, the 'y' in the third bracket, and the 'z' in the last bracket. Then all four brackets have the form $(1 + \text{something})$. You must therefore find the minimum value of

$$(1 + x)(1 + y/x)(1 + z/y)(1 + 16/z).$$

If you now put $x = \alpha$, $y/x = \beta$, $z/y = \gamma$, and $16/z = \delta$, then the expression takes the more symmetrical form

$$(1 + \alpha)(1 + \beta)(1 + \gamma)(1 + \delta).$$

However, α, β, γ, and δ are not 'independent' variables, since $\alpha \cdot \beta \cdot \gamma \cdot \delta =$ _____.

(3) You may still not know how to solve the problem, but at least it is now in a sufficiently nice form that you can look at some similar (but easier) cases to try to get some ideas. For example, you could start by trying to find the minimum value of $(1 + \alpha)(1 + \beta)$ given that $\alpha \cdot \beta = 4$.

Your *answer* to this easier problem should suggest a likely answer to the original problem (namely $\frac{1}{81}$). But your *method* of solving the easy problem will probably not work for the original problem. So take a closer look at the easy problem. First observe that $(1 + \alpha)(1 + \beta) = 1 + (\alpha + \beta) + \alpha\beta$. So if the value of $\alpha \cdot \beta$ is given, then you are really trying to minimize the value of $\alpha + \beta$. This is precisely the famous (and possibly familiar) problem of finding which rectangle with a prescribed area ($\alpha \cdot \beta$) has the smallest possible *e*i*e*e* ($2(\alpha + \beta)$). The minimum occurs when $\alpha = \beta$; that is, for a square. Can you use the same approach to find the minimum value of $(1 + \alpha)(1 + \beta)(1 + \gamma)$ given that $\alpha \cdot \beta \cdot \gamma = 8$?

(4) Another way of getting a feeling for the problem is to think in terms of graphs. Let $f(x) = x^2 + bx + c$ be a quadratic with negative roots $x = -\alpha$ and $x = -\beta$. Suppose that the constant term $c = \alpha\beta$ is fixed but that α and β can vary. Sketch the graph of $y = f(x)$. As α and β vary (but $c = \alpha\beta$ remains fixed), when does the value $f(1)$ take its minimum value? (That is, you want to minimize $1 + (\alpha + c/\alpha) + c$ as α varies.) What has this to do with the first easy case in (3) above?

Let $g(x) = x^3 + px^2 + qx + r$ be a cubic with three negative roots $x = -\alpha$, $x = -\beta$, and $x = -\gamma$. Suppose that the constant term $r = \alpha \cdot \beta \cdot \gamma$ is fixed but that α, β, and γ can vary. Sketch the graph of $y = g(x)$. As α, β, and γ vary, when does the value $g(1)$ take its minimum value? What has this to do with the second easy case in (3) above?

This graph idea may suggest a possible approach using calculus. (However, this approach is messy and slightly subtle.)

(5) For those who prefer simple algebra, the missing idea here, as in many

similar problems, is the 'AM–GM inequality'. (See 'A little useful mathematics' (Section 2.6) for a discussion of this result.) Use the AM–GM inequality for two variables to *prove* that if c is a constant and x_1 and x_2 satisfy $x_1 \cdot x_2 = c^2$, then

$$(1 + x_1)(1 + x_2) \geqslant (1 + c)^2,$$

with equality only if $x_1 = x_2 = c$.

Use the AM–GM inequality for three variables to prove that if c is a constant and x_1, x_2, and x_3 satisfy $x_1 x_2 x_3 = c^3$, then

$$(1 + x_1)(1 + x_2)(1 + x_3) \geqslant (1 + c)^3,$$

with equality only if $x_1 = x_2 = x_3 = c$.

Finally, use the AM–GM inequality for four variables to solve the original problem.

---

7. Let n and k be arbitrary positive integers. Prove that there exists a positive integer x such that $\frac{1}{2}x(x + 1) - k$ is divisible by 2^n.

---

(1) You may recognize $\frac{1}{2}x(x + 1)$ as the xth triangular number T_x, where

$$T_x = 1 + 2 + 3 + \ldots + x = \tfrac{1}{2}x(x + 1).$$

It is not clear whether this interpretation will help. But it seems that you have to show that

'for any given 2^n, there is a triangular number $T_x = \frac{1}{2}x(x + 1)$ to go with

each k, so that $\frac{1}{2}x(x + 1) - k$ is exactly divisible by 2^n'.

Is there another way of expressing this mouthful? What does it mean to say that '$\frac{1}{2}x(x + 1) - k$ is exactly divisible by 2^n'? What does that 'minus k' mean?

'$T - k$ is exactly divisible by t' means

'$T - k = t \times$ something (with no remainder)',

that is, '$T = t \times$ something $+ k$'.

If you think carefully about this last version, you will realize that k is the *remainder* when we divide T by t. So your problem boils down to showing that

'given any 2^n,

when you divide successive triangular numbers $T_x = \frac{1}{2}x(x + 1)$ by 2^n,

you get all possible remainders k'.

Explain why you only have to worry about values of $k \leqslant 2^n$. (Note that when $k = 2^n$ the 'remainder' would usually be taken to be 0. If we stick to *positive* k, as suggested in the question, then we cannot allow $k = 0$ as a remainder, but must take $k = 2^n$ in this case.)

(2) It is not clear where to go from here, but you are now well-placed to do a little experimenting, keeping an eye open for anything unexpected. Complete the following tables. Do you notice anything remarkable?

Let $n = 1$ k	1	2
first T_x with $T_x - k$ divisible by 2^1	$T_1 = 1$	$T_3 = 6$

Let $n = 2$ k	1	2	3	4
first T_x with $T_x - k$ divisible by 2^2	$T_1 = 1$	$T_3 = 6$	$T_2 = 3$	$T_7 = 28$

Let $n = 3$ k	1	2	3	4	5	6	7	8
first T_x with $T_x - k$ divisible by 2^3	$T_1 = 1$	$T_4 = 10$						

Let $n = 4$ k	1 2 3 4 5 6 7 8 9 10 11 12 13 14 15 16
first T_x with $T_x - k$ divisible by 2^4	

(3) The optimist might have hoped to obtain all possible remainders $1, 2, \ldots, 2^n$ using only the first 2^n triangular numbers. This idea breaks down right at the start:

When $n = 1$ $T_1 = 1$ and $T_2 = 3$ both correspond to remainder $k = 1$. We need to go on to $T_3 = 6$ to get remainder $k = 2$ (or, if you prefer, $k = 0$).

When $n = 2$ $T_1 = 1$, $T_2 = 3$, $T_3 = 6$ all leave *different* remainders when divided by 2^2. But $T_4 = 10$ leaves the same remainder as T_3. We need to go on to $T_7 = 28$ to get remainder $k = 4$ (or, if you prefer, $k = 0$).

When $n = 3$ we have to go all the way up to $T_{15} = 120$ to get remainder $k = 8$ (or, if you prefer, $k = 0$).

There really are many things that you should have noticed about the tables in (2). What is the largest value of x (or the largest triangular number T_x) that you need to use when $n = 4$? Which remainder does it correspond to? What would you expect to be the largest value of x (or triangular number T_x) you need to use when $n = 5$? Which remainder k do you expect will need this largest triangular number? Check your guesses by completing a similar table for $n = 5$.

(4) Once you notice what seems to be a general pattern you must formulate precisely what you believe to be true, and then try to *prove* that things really do work in the way you think they do. You know that you certainly cannot obtain all the remainders $k = 1, 2, \ldots, 2^n$ by looking at the first 2^n triangular numbers only. But it does look as though you can obtain every remainder by using at most $**i*e$ as many triangular numbers as this absolute minimum. In fact, it looks as though the first $2^{n+1} - 1$ triangular numbers will always be enough. If you looked more carefully at the 'repeats' in your tables, you may have noticed that each remainder k occurs *exactly twice*—and from two very closely related triangular numbers! Suppose that $\frac{1}{2}x(x + 1)$ and $\frac{1}{2}y(y + 1)$ leave the *same* remainder k when divided by 2^n. What does this tell you about x and y? (Factorize $x(x + 1) - y(y + 1)$.) How does this solve the original problem?

22nd British Mathematical Olympiad, 1986

---

1. Reduce the fraction N/D to its lowest terms when

$$N = 2\,244\,851\,485\,148\,514\,627, \qquad D = 8\,118\,811\,881\,188\,118\,000.$$

---

(1) You may be tempted to dismiss this problem as 'mere arithmetic'. It is certainly true that, although the problem is harder than it looks, elementary methods should succeed. However, even the most simple-minded approaches need to exploit one or two simple mathematical ideas (such as noticing that both N and D are divisible by $*i*e$, and that any number which consists of several repetitions of the same string of digits has an obvious factor). Whether you succeed in solving the problem this way, or get stuck, you should not be satisfied until you discover a more mathematical approach.

(2) So even if you think you know the answer, go back to the beginning and think things through again, step by step. Of the two numbers N and D, one looks far more promising as a starting point. Which one is it?

(3) The first thing to notice about D is that it has a factor $A = 1000$. In contrast, the numerator is divisible neither by 2, nor by 5. Hence no part of A can ever cancel. Thus, even in lowest terms, the denominator will still be a multiple of 1000.

  If you factorize $D = A \times D'$, then D' has an obvious four-digit factor $B = \underline{\quad\quad}$. And when you write

$$D = A \times B \times D'',$$

then D'' consists entirely of zeros and ones. Moreover, just as the repeated sequence of digits '8118' in D means that D has a factor '8118', so the repeated sequence '10001' in D'' means that D'' factorizes as $10\,001 \times \underline{\quad\quad}$.

Most interesting of all is the fact that, in this factorization of D'', the first factor $10\,001 = 10^4 + 1$, while the second factor can be written as _____.
(4) There is not much point in trying to break down these factors of D further *at this stage*—for it is not yet clear which bits of the denominator D have a chance of cancelling with the numerator N.

So turn your attention to N. The number N has something of the same structure as D, with three repeated blocks in the middle. (The repeating block '8514' stands out, but there is in fact another four-digit block that repeats three times. What is it?) You would like to exploit these repeating blocks. But how? There are two linked observations which might show how:

(a) First, if the numerator and denominator of the fraction N/D in lowest terms are going to be reasonably small, then it is essential that most, or perhaps all, of the factors $10^8 + 1$ and $10^4 + 1$ in D should cancel with N. Such factors arise (as you have seen) when a number has a block of digits which repeats after the appropriate number of steps (either eight or four). As it stands, N does not quite have this structure.

(b) However, although the numerator N appears to have a slightly different structure from that of D, that is only because the first three digits '224' and the last three digits '627' mess things up. If you increase the number formed by the three digits on the left-hand end, and decrease the number formed by the three digits on the right-hand end *by the same amount k*, then the number N is changed by $k \times$ _____ .

Notice that the second factor here is divisible by $10^8 + 1$! Moreover, there is one particular choice of k, namely $k =$ _____ , that produces a number

$$N' = N + k \times (10^{16} - 1)$$

which has exactly the same structure as D (so that N' can then be factorized in exactly the same way as D was in (3) above). What is this magic value of k? Use this to complete the solution very elegantly.

---

2. A circle S of radius R has two parallel tangents, t_1 and t_2. A circle S_1 of radius r_1 touches S and t_1; a circle S_2 of radius r_2 touches S and t_2; S_1 also touches S_2. All of the contacts are external. Calculate R in terms of r_1 and r_2.

---

(1) To get started you must draw a good diagram, showing all the information given, and marking in any points or lines which are likely to be important (even though they may not be explicitly mentioned in the problem).
(2) The fact that you are asked to 'calculate R in terms of r_1 and r_2' may seem slightly strange (since R is given first, the circle of radius r_1 is constructed next, and R and r_1 together determine r_2—so you might expect

to calculate r_2 in terms of R and r_1). Nevertheless, you should certainly expect there to be some kind of formula relating the three quantities R, r_1, and r_2; and such a formula can presumably then be written in the form '$R = \underline{\qquad}$'.

(3) The question is about three *i***e*. To specify a *i***e it is no good just marking the circumference: so you should have naturally marked the *e***e, and also any interesting *a*ii for each circle. Moreover, if the *e***e* of the circles are important, you should give them names: let C be the centre of the circle S, and let C_i be the centre of circle S_i $(i = 1,2)$. Since the question is about circles which are *tangent* to each other, you should naturally have drawn in the lines joining the centres of touching circles. And since the circles are tangent to two lines, you should mark the radii from the centres to the four points of tangency (and label the points so that you can refer to them: let P_i be the point at which S meets t_i $(i = 1,2)$). No other information is given, so it remains to interpret the diagram in some way which captures the fact that the three circles are tangent to each other and to the two parallel lines.

(4) The fact that the two lines (t_1 and t_2) are *e***e**i*u*a to the radii you have just marked should suggest that ****a*o*a*' theorem may come in handy. Calculate the distance d_1 from C_1 to the line P_1P_2, and the distance d_2 from C_2 to the line P_1P_2. Next, calculate the distance from C_1 to the line through C_2 parallel to t_2, and hence the distance d from C_2 to the line through C_1 parallel to P_1P_2. Finally, express the obvious relation between d_1, d_2, and d as an equation in R, r_1, and r_2.

(5) The relation between R, r_1, and r_2 is so unexpected and so beautiful that is is natural to ask whether there is not some simpler reason why the three quantities are related in this way. There are certainly many other solutions—for example, using angles and the cosine rule. You might also expect such a simple formula to have other interesting consequences. For example, consider the following. Let X_i be the point at which S_i touches t_i, and let X_iC_i meet S_i again at Y_i $(i = 1,2)$. What can you prove about the three points C, Y_1, and Y_2? And what interesting point lies on the line X_1X_2?

---

3. Prove that if m, n, and r are positive integers and

$$1 + m + n\sqrt{3} = (2 + \sqrt{3})^{2r-1},$$

then m is a perfect square.

---

(1) At first sight, there is no obvious reason whatsoever *why* the number m should always be a square, so it is natural to begin by calculating the first few odd powers: $(2 + \sqrt{3})^1 = \underline{\qquad}$, so $m = (\underline{\quad})^2$ is a square; $(2 + \sqrt{3})^3 = \underline{\qquad}$, so $m = (\underline{\quad})^2$ is a square; $(2 + \sqrt{3})^5 = (2 + \sqrt{3})^3(2 + \sqrt{3})^2 = \underline{\qquad}$, so $m = (\underline{\quad})^2$ is a square; $(2 + \sqrt{3})^7 = (2 + \sqrt{3})^5(2 + \sqrt{3})^2 = \underline{\qquad}$, so

$m = (\underline{\quad})^2$ is a square. Such calculations may convince you that there is indeed something to prove; but the numbers quickly become uncomfortably large, and do not immediately suggest any obvious pattern. It may even be worth pointing out that the actual *numbers* which drop out of such calculations can often *prevent* you from seeing how to solve the problem: for example, the sequence $1, 5, 19, 71, \ldots$ is so unexpected that it is tempting to think that the fact that m is always a square may be purely accidental. Instead of calculating more numbers, what you need is some insight into the way in which the value of m is determined *in general*.

   (While you are still feeling your way into the problem, it is natural to ask *why* the question is restricted to *odd* powers $(2 + \sqrt{3})^{2r-1}$. In such cases it is always worth checking a few examples to see whether *even* powers really do behave differently: $(2 + \sqrt{3})^0 = 1$, so $m = 0$, which is a square; $(2 + \sqrt{3})^2 =$ _____, so $m = $ _____, which is not a square; $(2 + \sqrt{3})^4 = (7 + 4\sqrt{3})^2 =$ _____, so $m = $ _____, which is not a square.)

(2) It is natural to hope that calculating the odd powers with $r = 1$, $r = 2$, $r = 3$, and $r = 4$ will not only show that m is a square, but provide a clue as to *what m is the square of* when $r = 5$, $r = 6$, and so on. Sadly, that is rather unlikely in this case! So you have no choice but to try to think about the general case. Choose a general value of r and let

$$(2 + \sqrt{3})^{2r-1} = a + b\sqrt{3}.$$

You want to prove that $a - 1$ is always a square. Use the $*i*o*ia*$ theorem to write down the sum of all the terms in the expansion of $(2 + \sqrt{3})^{2r-1}$ which do not include $\sqrt{3}$, and which therefore have sum equal to a. All of these terms except one involve an $e*e*$ power of $\sqrt{3}$; and so are multiples of 3; the one remaining term is a power of 2, so is *not* a multiple of 3. Hence a is never a multiple of 3. It follows that, for each choice of r, either $a + 1$ or $a - 1$ is always a multiple of 3. Is it always $a + 1$ that is a multiple of 3? Or is it always $a - 1$? Or is it sometimes one and sometimes the other? Prove your claim.

(3) To solve any good Olympiad problem you need an idea. $\sqrt{3}$ is defined to be the *positive* square root of 3; that is, $\sqrt{3}$ is the positive root of the equation $x^2 = 3$. In multiplying out the expression $(2 + \sqrt{3})^{2r-1}$, you repeatedly used the fact that $(\sqrt{3})^2 = 3$ to simplify and collect up terms to obtain the two coefficients a and b. The equation $x^2 = 3$ has another root, namely _____, which behaves exactly like $\sqrt{3}$ (in that $(\underline{\quad})^2 = 3$). This means that the two expressions $(2 + \sqrt{3})^{2r-1}$ and $(2 - \sqrt{3})^{2r-1}$ are so similar that, if you let

$$(2 - \sqrt{3})^{2r-1} = a' + b'\sqrt{3},$$

you should be able to say straight away (without doing any messy calculations) how a and a' are related, and how b and b' are related. Use this, together with the observation that $(2 + \sqrt{3}) \cdot (2 - \sqrt{3}) = $ _____ , to prove that $a^2 - 3b^2 = 1$.

(4) Given what you found at the end of (2), it should be almost impossible to resist the temptation to take the 1 to the other side and to rewrite this equation in the form (_____)(_____) $= 3b^2$. From (2) you know that $a + 1$ is always a multiple of 3; it follows that $a - 1$ and $(a + 1)/3$ are integers, and that their product is a perfect square. All you need to finish off the question is to prove that $HCF(a - 1, (a + 1)/3) = $ ____ (see 'A little useful mathematics', Section (4.3.3)).

[Alternatively, you might spot a recurrence relation for the sequence $A_1 = 1, A_2 = 5, A_3 = 19, A_4 = 71, \ldots$, which appears to determine how each term can be calculated from the previous two terms. You then have two separate sequences:

- the original sequence $a_1, a_2, a_3, \ldots$, where a_r is the term independent of $\sqrt{3}$ in the expansion of $(2 + \sqrt{3})^{2r-1}$, and
- the sequence A_r, defined by a simple recurrence relation.

You may then be able to (i) prove a recurrence for the sequence (a_r), and (ii) to prove (by induction) that $a_r - 1 = A_r^2$ for every $r \geqslant 1$.]

---

4. Find, with proof, the largest real number K (independent of a, b, and c) such that the inequality

$$a^2 + b^2 + c^2 > K(a + b + c)^2$$

holds for the lengths a, b, and c of the sides of any *obtuse-angled* triangle.

---

(1) It is not easy to grasp what exactly is being asked. If you only had to find a value of K such that $a^2 + b^2 + c^2 > K(a + b + c)^2$ always holds, then you could choose $K = 0$ (or any negative value of K). Thus, it is important to find the *best possible* value of K—which in this case means *the largest possible K*. But that is not all, for the numbers a, b, and c are not just any old numbers! You have to find the *largest possible* value of K such that the inequality holds whenever the numbers a, b, and c are the *sides of an obtuse-angled triangle*.

(2) Explain why K must be $\leqslant 1$.

(3) For the proof of (2) you don't need to worry about obtuse-angled triangles; you only need the fact that a, b, and c are *o*i*i*e. To understand the problem as given, it may help to start by finding what the best possible value of K would be if a, b, and c were not restricted to being the sides of an obtuse-angled triangle, but could be any three positive real numbers. The answer to this question exploits the standard facts that $(a - b)^2$, $(b - c)^2$, and $(c - a)^2$ are all $\geqslant 0$; hence their sum $(a - b)^2 + (b - c)^2 + (c - a)^2 \geqslant 0$ (with

equality holding precisely when $a = b = c$). Rearranging this gives $a^2 + b^2 + c^2 \geqslant (a + b + c)^2/3$, with equality holding precisely when $a = b = c$. This has two interesting consequences.

(3A) The first consequence may be slightly unexpected. You have proved that, if a, b, and c are positive, then $a^2 + b^2 + c^2 \geqslant (a + b + c)^2/3$. It follows that $a^2 + b^2 + c^2 > K(a + b + c)^2$ whenever $K < \frac{1}{3}$, but *not* when $K = \frac{1}{3}$ (since the inequality with $K = \frac{1}{3}$ becomes an equality whenever $a = b = c$). Hence there is no *largest* value of K such that '$a^2 + b^2 + c^2 > K(a + b + c)^2$' holds for *all* triples a, b, and c of positive real numbers!

(3B) The second consequence is that, if a, b, and c are the lengths of the sides of an obtuse-angled triangle, then you know that $a = b = c$ is impossible (Why?), so the strict inequality $a^2 + b^2 + c^2 > \frac{1}{3}(a + b + c)^2$ certainly holds for all such triples a, b, and c. Hence the best possible value of K is $\geqslant \frac{1}{3}$.

(4) You now know that the required value K satisfies $\frac{1}{3} \leqslant K \leqslant 1$. Suppose that you choose your favourite obtuse-angled triangle, and calculate the ratio

$$\frac{(a^2 + b^2 + c^2)}{(a + b + c)^2}.$$

You have to find the largest value of K for which this ratio is $> K$ for *all* obtuse-angled triangles. Hence the ratio must be $> K$ for each individual obtuse-angled triangle. So the ratio that you obtain for any particular obtuse-angled triangle provides an *upper bound* for the required value of K. For example, if you take the triangle with sides of length 1, 1, and $\sqrt{3}$ (having angles 30°, 30°, and 120°), then the ratio is equal to $5/(2 + \sqrt{3})^2 = 5(2 - \sqrt{3})^2 = 35 - 20\sqrt{3}$. Prove (without using a calculator) that $35 - 20\sqrt{3} > \frac{1}{3}$. Conclude that $35 - 20\sqrt{3} \geqslant K \geqslant \frac{1}{3}$.

(5) It should now be clear what the problem is asking, even if you still do not know how to solve it. However, you have already seen the possible advantage of rearranging the given inequality with all the a's, b's, and c's on one side, with the elusive 'K' on the other. You have to find the largest value of K such that

$(*)$
$$\frac{(a^2 + b^2 + c^2)}{(a + b + c)^2} > K$$

holds whenever a, b, and c are the sides of an obtuse-angled triangle. Once the LHS is in this form, one of the first things you might notice is that when a given obtuse-angled triangle is enlarged—say with scale factor λ—the expression on the LHS is unchanged! So, given any obtuse-angled triangle, you may enlarge (or shrink) the triangle in any way that happens to be convenient. In particular, you may enlarge the given triangle so that the perimeter $a + b + c$ is as simple as it could possibly be; namely, equal to ___. The LHS of the inequality $(*)$ then becomes much simpler, and all you have to find is the largest possible value of K, such that

$$a^2 + b^2 + c^2 > K$$

holds whenever a, b, and c are the sides of an obtuse-angled triangle *with perimeter $a + b + c =$ ___* .

(6) In any obtuse-angled triangle, one side is definitely longer than the other two: let c be the length of the longest side. It is then natural to use the $*o*i*e$ rule to obtain a formula expressing c^2 in terms of a, b, and the angle C. Do this, and use this expression for c^2 to obtain an expression for '$a^2 + b^2 + c^2$'.

At the same time, you should not forget that $c = 1 - (a + b)$. Combine this with the expression for c^2 given by the cosine rule to prove the important relation

$$2(a + b) - 2ab(1 + \cos C) = 1.$$

It is much harder to handle the algebra if all three variables a, b, and c are allowed to vary at the same time. One way of simplifying things which is often useful is to concentrate on one permissible value of c at a time; that is, to keep c fixed for the moment. Then $a + b = 1 - c$ is also fixed; and the equation that you have just proved then shows that $2ab(1 + \cos C)$ is also fixed. Since you are trying to find a universal lower bound for the expression '$a^2 + b^2 + c^2$', it is natural, for each value of c, to rewrite this expression in a form which suggests a natural lower bound: for example, involving squares (which are always $\geqslant 0$), and terms depending only on $a + b$ and $2ab(1 + \cos C)$ (which remain fixed when c is fixed). Use $c = 1 - (a + b)$ to show that

$$a^2 + b^2 + c^2 = \frac{(a + b)^2}{2} + \frac{(a - b)^2}{2} + (a + b)^2 - 2ab(1 + \cos C).$$

It follows, for each permissible value of c, that

$$a^2 + b^2 + c^2 \geqslant \frac{3(a + b)^2}{2} - 2ab(1 + \cos C) = \frac{3(a + b)^2}{2} + 1 - 2(a + b),$$

with equality holding if and only if $a = b$ (that is, if and only if the obtuse-angled triangle is $i*o**e*e*$). Hence, for a given value of c, the best possible lower bound $K(c)$ is $2a^2 + 2a^2(1 - \cos C) = \frac{1}{2}(1 - c)^2 + c^2$.

It remains to find the best value of K such that $2a^2 + 2a^2(1 - \cos C) > K$ whenever the angle C is $o**u*e$. For this, it is sufficient to concentrate on the case which gives the smallest value of $a^2 + b^2 + c^2$ for each c; namely, when $a = b$ (so that the triangle is isosceles). Show that $c > \sqrt{2} \cdot a$, whence $c >$ ___ , giving $K(c) = \frac{1}{2}(1 - c)^2 + c^2 >$ ___ , and this is the required value of K (since c can be arbitrarily close to $\sqrt{2} - 1$).

[This problem is genuinely hard! I have tried to avoid presenting a slick, but completely unmotivated, solution. There are other approaches, but they all have to face the subtlety that the extremal example (an isosceles right-angled triangle) is not 'obtuse-angled', and so does not belong to the given domain. One way of avoiding the cosine rule is to let c ($\geqslant a, b$) be fixed.

Then $a^2 + b^2 = [(a + b)^2 + (a - b)^2]/2 \geqslant (a + b)^2/2 = (1 - c)^2/2$, since $a + b + c =$ _____ . Hence $a^2 + b^2 + c^2 \geqslant c^2 + (1 - c)^2/2$, with equality if and only if $a = b$.]

---

5. Find, with proof, the number of *permutations*

$$a_1, a_2, a_3, \ldots, a_n \quad \text{of} \quad 1, 2, 3, \ldots, n$$

such that

$$a_r < a_{r+2} \quad \text{for } 1 \leqslant r \leqslant n - 2 \quad \text{and} \quad a_r < a_{r+3} \quad \text{for } 1 \leqslant r \leqslant n - 3.$$

---

(1) A reasonable way of trying to obtain a feeling for what the question means is to work with small values of n in turn, and to list, and hence count, all permutations satisfying the given conditions. Let P_n denote the number of permutations of $1, 2, \ldots, n$ which satisfy the given conditions.

When $n = 1$ or $n = 2$ the conditions do not apply (Why not?), so P_n counts *all* permutations: hence $P_1 = 1$ and $P_2 =$ __ . When $n = 3$, the second condition does not apply (Why not?), and the first condition requires only that $a_1 < a_3$; hence, either $a_3 = 2$ (so $a_1 =$ __ , and there is just one such permutation), or $a_3 = 3$ (in which case we must have either $a_1 =$ __ or $a_1 =$ __ , so there are exactly __ such permutations). Thus $P_3 =$ __ . Now calculate P_4.

(2) While it can be helpful to work your way into a problem like this, there are two major snags. The most frequent (but least significant) snag, is that it is easy to make a mistake! The data that you generate will then *mislead* rather than help. (This is one instance of the infamous 'GIGO' (*Garbage In, Garbage Out*) principle: if your basic data is flawed, your conclusions are doomed.) The more serious snag is that you have to solve the problem for *all* n, and, although you may sometimes be lucky, you should not assume that the numbers you calculate for small values of n will necessarily give you a clue as to what happens in general. In this instance, for example, P_4 does not have the 'obvious' value which may be suggested by the values of P_1, P_2, and P_3. But if you work out P_4 and P_5, you should come up with another idea. That will at least give you something to aim at.

(3) But whether or not you think you know what the answer is, you still have to find some way of counting which will give the value of P_n for *every* $n \geqslant 1$. One natural approach in problems such as this one is to try to find a *e*u**e**e *e*a*io*; that is, you want to express the value of P_n in terms of P_{n-1}, P_{n-2}, and so on.

One thing that you should have noticed when calculating P_3, P_4, and P_5 was that the largest available number n always has to go in one of two places. What are these two places? Use the given conditions '$a_r < a_{r+2}$' and '$a_r < a_{r+3}$' to *prove* that this restriction always applies.

(4) The hint in (3) should have been enough to show the way; for once you know that the number n has to go in either the very last position or the

last-but-one position, the stage is set for a recurrence relation. Suppose that n goes in the very last position; then the first $n-1$ terms form a permutation of the numbers $1, 2, \ldots, n-1$ which satisfies the standard conditions '$a_r <$ a_{r+2}' and '$a_r < a_{r+3}$', and there are exactly ____ such permutations. Suppose, on the other hand, that n goes in the last-but-one position; then the standard conditions guarantee that the last position must be occupied by ____, so the last two positions are uniquely determined and the first $n-2$ terms form a permutation of $1, 2, \ldots, n-2$ which satisfies the standard conditions, so there are exactly ____ such permutations. Hence $P_n = P_{n-1} + P_{n-2}$. Finally, since the first two terms are the second and third *i*o*a**i numbers, it follows that $P_n = F_{n+1}$ for all $n \geqslant 1$.

---

6. *AB*, *AC*, and *AD* are three edges of a cube. *AC* is produced to *E* so that $AE = 2AC$, and *AD* is produced to *F* so that $AF = 3AD$. Prove that the area of the section of the cube by any plane parallel to *BCD* is equal to the area of the section of tetrahedron *ABEF* by the same plane.

---

(1) Let the cube have side 1, and let A', B', C', and D' be the vertices of the cube diametrically opposite A, B, C, and D respectively. Since the edges AB, AC, and AD belong to both the cube and the tetrahedron $ABEF$, it is natural to take A as 'origin', and to position the cube in the positive octant, with AB, AC, and AD as x-, y-, and z-axes.

(2) The plane BCD has equation $x +$ $y + z =$ ____ . Each plane parallel to BCD has an equation of the same form, $x +$ $y + z = c$, with different constants c on the RHS.

When $c < 0$, such a plane cuts neither the cube nor the tetrahedron $ABEF$. And when $0 \leqslant c \leqslant 1$, the plane $x + y + z = c$ cuts both the cube and the tetrahedron in exactly the same triangle (parallel to BCD, and with area $c^2 \cdot$ area(BCD)). Hence the two cross-sections have equal area whenever $c \leqslant 1$.

The plane $x + y + z = 3$ cuts the cube in the single point ___, and cuts the tetrahedron $ABEF$ in the single point ___. When $c > 3$, the plane $x + y + z = c$ cuts neither the cube nor the tetrahedron.

You therefore need only consider the remaining cases in which $1 < c < 3$.

(3) The plane $x + y + z = 2$ cuts the cube in the **ia***e _____ , and cuts the tetrahedron $ABEF$ in the **ia***e PEQ, where P is the midpoint of BF and Q is the midpoint of DF. Hence, when $c \geqslant 2$, the only part of the cube that need concern you is the tetrahedron $B'C'D'A'$, and the only part of the

tetrahedron *ABEF* that need concern you is the tetrahedron *PEQF*. However, the bases *B'C'D'* and *PEQ* of these two tetrahedra lie in the same plane $x + y + z = 2$; and the two tetrahedra have equal 'height' perpendicular to this plane (since the parallel plane $x + y + z = 3$ passes through both apexes *A'* and *F*). Show that the two bases *B'C'D'* and *PEQ* have equal areas. Hence prove that, for each value of c, $2 \leqslant c \leqslant 3$, the cross-sections produced by the plane $x + y + z = c$ with the tetrahedra *B'C'D'A'* and *PEQF* have equal areas.

(4) It remains to sort out what happens for values of c between 1 and 2. Let $c = 1 + t$, where $0 \leqslant t \leqslant 1$.

(4A) Consider first the cross-section produced by the intersection of the plane $x + y + z = 1 + t$ with the cube. This cross-section is equal to the triangle *BCD* when $t =$ ___, and is equal to the triangle *B'C'D'* when $t =$ ___. For values of t between 0 and 1, the plane cuts the cube between *BCD* and *B'C'D'* in a *e*a*o*, having successive sides parallel to *B'C'*, *DB*, *C'D'*, *BC*, *D'B'*, and *CD*. Hence the angles of the hexagon are all equal to ___°.

If the plane $x + y + z = 1 + t$ cuts *DC'* at *X* and *DB'* at *Y*, then $DX = DY =$ ___. Hence $XY =$ ___. Similarly, if the same plane cuts *C'B* at *Z*, then $C'X = C'Z =$ ___, so $XZ =$ ___. Since $BZ =$ ___, it follows that the next side of the hexagon has the same length as ___, and that the edge

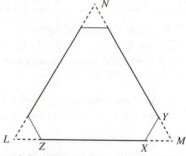

lengths of the cross-section are alternately $t\sqrt{2}$ and $(1 - t)\sqrt{2}$. Hence the hexagon is obtained from an equilateral triangle *LMN* of side $XZ + 2 \cdot XY = (1 + t)\sqrt{2}$ by chopping off three congruent corners of side length $XY =$ ___. If you wish, it is now a simple matter to calculate the area of this cross-section (although it may be better to wait and see whether this is strictly necessary).

(4B) Consider next the cross-section produced by the intersection of the plane $x + y + z = 1 + t$ with the tetrahedron *ABEF*. When $0 \leqslant t \leqslant 1$, this cross-section is a *ua**i*a*e*a* with successive sides parallel to *BD*, *DC*, *CB*, and *PE*. Moreover:

(i) the plane cuts the face *AEF* in precisely the segment *MN*;

(ii) the plane cuts the face *ABF* in a segment *NS*, where *S* lies on *NL*;

(iii) the plane cuts the face *ABE* in a segment *MR*, where *R* lies on *ML*.

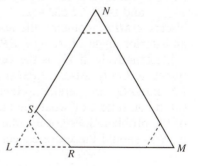

Thus you only need to calculate the lengths *NS* and *MR*.

In the plane $y = 0$, the line NL has equation $x + z =$ ____ . Write down the equation of the line BF. Hence find the x co-ordinate $x(S)$ of the point S, and show that $NS = x(S) \cdot \sqrt{2}$.

In the plane $z = 0$, the line ML has equation $x + y =$ ____ . Write down the equation of the line BE. Hence find the x co-ordinate $x(R)$ of the point R, and show that $MR = x(R) \cdot \sqrt{2}$.

Finally, show that $LR = 2t \cdot \sqrt{2}$, $LS = 3t\sqrt{2}/2$, so that the triangle LRS has area equal to $3 \times$ area($\triangle XYN$).

[Alternatively, you might prefer to work out the 3-D co-ordinates of L, R, and S, and use these to calculate area($\triangle LRS$).]

21st British Mathematical Olympiad, 1985

1. Circles S_1 and S_2 both touch a straight line p at the point P. All points of S_2, except P, are in the interior of S_1. The straight line q: (i) is perpendicular to p; (ii) touches S_2 at R; (iii) cuts p at L; and (iv) cuts S_1 at N and M, where M is between L and R.
   (a) Prove that RP bisects $\angle MPN$.
   (b) If MP bisects $\angle RPL$, find, with proof, the ratio of the areas of the circles S_1 and S_2.

(1) Draw a decent diagram.
(2) The question is all about *i***e* (S_1 and S_2) and *a**e*** (p and q), so you should have already started to apply the two most basic facts that you know:

(a) the two tangents LP and LR from L to S_2 must have the same *e**** (so triangle LPR is i*o**e*e*); and

(b) the 'alternate segment theorem' (which states that the angle between a tangent and a chord of the same circle is equal to the angle subtended in the opposite segment).

The first of these facts shows that $\angle LPR = \angle LRP$. And if we denote the point at which PN cuts S_2 by X, then the second fact shows that $\angle NRX =$ $\angle$ ____ and that $\angle LPM = \angle$ ____ . You should now be able to prove that RP bisects $\angle MPN$ (combine the above observations with the fact that $\angle LRP$ is an exterior angle of triangle RPN).

[Alternatively, if O_2 is the centre of S_2, then $PLRO_2$ is a **ua*e, so PR bisects $\angle LPO_2$. Also $\angle LPM = \angle PNL$ (by (b)), and $\angle PNL = \angle NPO_2$ (since LN and PO_2 are parallel). Hence $\angle MPR = \angle RPN$.]
(3) If you reflect on what you have just proved, it should be clear that part (a) of the problem does not use the given information that 'q is perpendicular to p'. You should therefore expect this to play an important role in part (b).

(4) Now suppose that *MP* also bisects ∠ *RPL*. You have to calculate the ratio of the areas of S_1 and S_2. In other words, you have to find the ratio of the **ua*e of the radii of the two circles, so you need to mark the *e****e* O_1 and O_2 of the two circles S_1 and S_2. These two centres lie on the perpendicular to the tangent *p* through the point *P*.

What can you say about the quadrilateral $O_2 RLP$? What does this tell you about ∠ *LPR*? What can you say about ∠ *MPN*? And what does this tell you about ∠ $MO_1 N$? So what can you conclude about the lines $O_2 L$ and $O_1 M$? Finally, what does this prove about the ratio $O_1 M : O_2 R$—and hence about $(O_1 M)^2 : (O_2 R)^2$?

---

2. Given any three numbers *a*, *b*, and *c* between 0 and 1, prove that not all of the expressions $a(1-b)$, $b(1-c)$, and $c(1-a)$ can be greater than $\frac{1}{4}$.

---

(1) At first sight, the wording may seem strange: you have to 'prove that not all of the expressions $a(1-b)$, $b(1-c)$, and $c(1-a)$ can be greater than $\frac{1}{4}$', but you have no idea which one is supposed to have this property! However, since you are only told that '*a*, *b*, and *c* lie between 0 and 1', there is no reason why any particular one of the three expressions should be $\leqslant \frac{1}{4}$: if $a = \frac{1}{2}$ and $b = c = \frac{1}{4}$, then $b(1-c) \leqslant \frac{1}{4}$; whereas if $a = b = \frac{1}{4}$ and $c = \frac{1}{2}$, then it is $a(1-b)$ that is $\leqslant \frac{1}{4}$.

(2) There are several approaches which work here, but they all come down to the same basic fact: if '*a* lies between 0 and 1', then you should know (or be able to prove for yourself) a fundamental inequality of the form '$a(1-a) \leqslant$ ___ '. [Given any two positive real numbers *x* and *y*, the AM–GM inequality guarantees that $\sqrt{xy} \leqslant (x+y)/2$; now put $x = a$ and $y = (1-a)$.]

(3) You have to prove that *not all* of $a(1-b)$, $b(1-c)$, and $c(1-a)$ can be greater than $\frac{1}{4}$ *at the same time*. (You have already seen that each one can be $> \frac{1}{4}$, provided that the other two are small.) One way to prove that something *must* happen is to show that it could not possibly be *false*. It therefore makes sense to turn the question around and to ask whether all three of these expressions could possibly be $> \frac{1}{4}$ *at the same time*. If that were to happen, then the **o*u** of the three expressions

$$a(1-b) \cdot b(1-c) \cdot c(1-a)$$

would be $> (\frac{1}{4})^3$. Now rearrange the six factors and use the inequality you proved in (2) to show that the above product is always $\leqslant (\frac{1}{4})^3$. Hence complete the solution.

---

3. Given any two non-negative integers *n* and *m* with $n \geqslant m$, prove that

$$\binom{n}{m} + 2\binom{n-1}{m} + 3\binom{n-2}{m} + \ldots + (n+1-m)\binom{m}{m} = \binom{n+2}{m+2}.$$

(1) If you want to solve any mathematical problem, you need to know a thing or two to get started—although you can usually manage with much less than you think! The binomial coefficients on the LHS of the identity to be proved correspond to entries in Pascal's triangle *with fixed second 'co-ordinate' m*. Thus these

```
              1
           1     1
         1    2    1
       1    3    3    1
     1    4    6    4    1
   1    5   10   10   5    1
 1    .    .    .    .    .    1
```

entries lie on a 'diagonal' which starts at $\binom{m}{m}$ and which slopes down to the left—like the diagonal picked out in bold in the figure here. Notice that $1 + 2 + 3 = \underline{\quad}$, and that $1 + 2 + 3 + 4 = \underline{\quad}$. This should suggest that it might be helpful to begin by proving the identity

$$(*) \qquad \binom{n}{m} + \binom{n-1}{m} + \binom{n-2}{m} + \ldots + \binom{m}{m} = \binom{n+1}{m+1}.$$

Then rewrite the LHS of the identity to be proved in the form

$$\binom{n}{m} + \binom{n-1}{m} + \binom{n-2}{m} + \ldots + \binom{m}{m}$$
$$+ \binom{n-1}{m} + \binom{n-2}{m} + \ldots + \binom{m}{m}$$
$$+ \binom{n-2}{m} + \ldots + \binom{m}{m}$$
$$\cdots\cdots$$
$$+ \binom{m}{m}$$

and use (*) over and over again to obtain a sum like the LHS of (*), but with n and m replaced by $n + 1$ and $m + 1$. Hence complete the solution.

(2) Another approach is to notice that the RHS of the identity to be proved counts the number of subsets of the set $\{1, 2, \ldots, n + 2\}$ which have $m + 2$ elements. The first term of the LHS clearly counts the number of such subsets which include both $n + 1$ and $n + 2$. Explain why the second term on the LHS counts those subsets which contain n and just one of $n + 1$ and $n + 2$. Continue in this way to explain how the LHS of the required identity accounts for all possible subsets of the set $\{1, 2, \ldots, n + 2\}$ which have $m + 2$ elements.

---

4. The sequence f_n is defined by $f_0 = 1$ and $f_1 = c$, where c is a positive integer, and for all $n \geqslant 1$ by

$$f_n = 2f_{n-1} - f_{n-2} + 2.$$

Prove that, for each $k \geqslant 0$, there exists h such that $f_k f_{k+1} = f_h$.

(1) Later questions on the BMO tend to be harder than earlier questions! As it stands, this one may seem thoroughly opaque. You are not told the value of c; the recurrence relation is unfamiliar; and you are asked to prove that, 'for each $k \geqslant 0$, there exists an h' without having any idea what the value of h may be.

(2) One way to escape from this feeling of helplessness is to choose a simple value of c and to calculate a few terms of the sequence, while keeping your wits about you to try to see what is going on. The simplest 'positive integer' to try for c is presumably $f_1 = c = \underline{\quad}$. So try this value for c, and calculate the first 18 or so values:

$$f_0 = 1, \qquad f_1 = c = 1, \qquad f_2 = 3, \qquad f_3 = 7, \qquad f_4 = 13,$$
$$f_5 = 21, \qquad f_6 = 31, \qquad f_7 = \underline{\quad}, \qquad f_8 = \underline{\quad}, \qquad f_9 = \underline{\quad},$$
$$f_{10} = \underline{\quad}, \qquad f_{11} = \underline{\quad}, \qquad f_{12} = \underline{\quad}, \qquad f_{13} = \underline{\quad}, \qquad f_{14} = \underline{\quad},$$
$$f_{15} = \underline{\quad}, \qquad f_{16} = \underline{\quad}, \qquad f_{17} = \underline{\quad}, \qquad f_{18} = \underline{\quad}.$$

Now look at $f_1 \cdot f_2 = \underline{\quad} = f_?$, $f_2 \cdot f_3 = \underline{\quad} = f_?$. Remember that you are trying to *predict* the value of h for each value of k; so after calculating the first few products, risk a guess about what seems to be going on and make a prediction before you calculate the next product $f_3 \cdot f_4 = \underline{\quad} = f_?$.

(3) Now repeat all of this with $c = 2$, and try to understand how the new value of c affects the value of h. When you think you understand, predict what you expect to happen when $c = 3$, and check a few cases.

(4) You are probably now in a position to formulate a conjecture which expresses the h in '$f_k \cdot f_{k+1} = f_h$' in terms of k and c. If you try to prove this conjecture directly using induction, you may have problems: while it is easy to check your formula for $k = 0$, the induction step is not at all easy. You may assume that $f_{k-1} \cdot f_k$ has the form that you expect, and then use the given recurrence multiplied by f_k,

$$f_k \cdot f_{k+1} = 2f_k^2 - f_{k-1} \cdot f_k + 2f_k,$$

to try to prove that the same formula holds for $f_k \cdot f_{k+1}$, but it is not clear what to do next—*unless you could somehow discover a formula for each term f_k in terms of k and c.* (There may be other ways to solve the problem, but this would seem to be the most direct approach.)

(a) When $c = 1$, you should notice that each f_k is slightly less than k^2; it is then not hard to guess the formula $f_k = k^2 - \underline{\qquad}$.

(b) When $c = 2$, things are even easier, since then each $f_k = \underline{\qquad}$.

(c) When $c = 3$, the previous two cases soon suggest trying $f_k = \underline{\qquad}$, and this should lead you to conjecture a general formula for f_k in terms of k and c.

This formula can be proved relatively easily (by induction) using the given recurrence. Your conjectured value of h can then be checked by direct algebraic computation.

---

5. A circular hoop of radius 4 cm is held fixed in a horizontal plane. A cylinder with radius 4 cm and length 6 cm rests on the hoop with its axis horizontal, and with each of its two circular ends touching the hoop at two points. The cylinder is free to move subject to the condition that each of its circular ends always touches the hoop at two points. Find, with proof, the locus of the centre of one of the cylinder's circular ends.

---

(1) Draw a diagram showing the initial position of the hoop and the cylinder (that is, with the axis of the cylinder horizontal). Then draw diagrams showing possible subsequent positions. What is the maximum possible angle of tilt of the axis of the cylinder?

(2) Let C be the centre of one of the cylinder's circular ends. You have to find the locus of C. Finding loci, or equations of curves and surfaces, does not normally require anything more sophisticated than Pythagoras' Theorem. In this case, the problem is made harder by the fact that you are not told what point to use as origin, or what lines to use as axes. It is also unclear what sort of locus to expect.

Any acceptable position of the cylinder can be rotated around the vertical line through the centre of the hoop, so the required locus will be a three-dimensional *u**a*e. Given the rotational symmetry of this surface, it is natural to choose the vertical line through the centre of the hoop as the z-axis. Then, for each fixed value of z, the locus of C will be a *i***e. Hence, it will suffice if you can describe the locus in any fixed vertical plane containing the z-axis (since you can then rotate this part of the locus around the z-axis). So, fix a y direction (to the right, say), let C be the centre of the cylinder's right-hand circular end, and restrict C to move in the y–z plane.

In the absence of any other clues, it makes sense to choose the centre O of the hoop as origin of co-ordinates, with the radius to the right as the y-axis, meeting the hoop at the point Y:

(a) What are the y and z co-ordinates of the point Y?

(b) Find the co-ordinates (y, z) of C when the axis of the cylinder is horizontal.

(c) Now tilt the cylinder upwards so that the right-hand circular end touches the hoop in two 'coincident' points (at $Y = (4,0)$). Find the co-ordinates of C. (This is completely elementary. However, you will need a combination of (i) persistence, and (ii) faith that 3-D diagrams can almost always be analyzed using only Pythagoras and similar right-angled triangles.)

(d) Finally, tilt the cylinder downwards until the left-hand circular end touches the hoop in two coincident points (at $(-4, 0)$). Find the co-ordinates of C. (This should be straightforward once you have done part (c). But be prepared for a surprise!)

(3) For a general solution you could find the co-ordinates (y, z) of the point C when the axis of the cylinder is inclined at angle θ to the horizontal, and hence find the equation of the locus. An alternative approach is to think about the results of your calculations in parts (b)–(d) of (2) above. What is the most likely possibility for a curve which passes through the point $(3, 3)$, and which passes through the two points $(\frac{72}{25}, \frac{54}{25})$ and $(\frac{72}{25}, \frac{96}{25})$? The symmetry of these three points about the line $z = 3$ should lead you to check the distances of these three points from the point (____, ____). You should now have a fairly clear idea of what the locus of C must be, and can now try to show that a general point on this curve between the two extreme points found in parts (c) and (d) of (2) is indeed the position of C for some inclination of the cylinder's axis.

[Alternatively, you may now realize that it suffices to explain why the midpoint of the axis of the cylinder remains fixed throughout the motion!]

---

6. Show that the equation $x^2 + y^2 = z^5 + z$ has infinitely many solutions in positive integers x, y, and z having no common factor greater than 1.

---

(1) What is the 'simplest' possible triple x, y, z (having no common factors) which satisfies $x^2 + y^2 = z^5 + z$? What is the next simplest solution? Make a list of fifth powers and use it to find the *next* simplest solution.

(2) It makes sense to interpret 'simplest' as meaning 'with smallest possible value of z'. $z = 1$ certainly works (with $x =$ ____ and $y =$ ____); $z = 2$ also works (with $x =$ ____ and $y =$ ____). And though $z = 3$ and $z = 4$ do not correspond to solutions, you probably discovered that $z = 5$ does work (with $x =$ ____ and $y =$ ____).

At this stage all that you know is that $z = 1$, $z = 2$, and $z = 5$ correspond to solutions, but $z = 3$ and $z = 4$ do not. For a quick solution, you need to notice two things. Observe first that the simplest solution of all is a little special (since any triple with $x = y = z$ would normally contradict the requirement that x, y, and z have no common factors). Thus we should concentrate on the fact that $z = 2$ and $z = 5$ work (and on the fact that $z = 3$ and $z = 4$ do not). Notice next that the values of z which work—namely 2 and 5—are precisely those which can themselves be written as the sum of two non-zero squares (and the values which do not work—namely 3 and 4—are precisely those which cannot be written as the sum of two non-zero squares). Once you have made this jump, the rest is fairly straightforward:

(a) Show that, if a, b, c, and d are integers, then the product $(a^2 + b^2)(c^2 + d^2)$ can always be rewritten as the sum of two squares. (If $z = a + bi$ and $w = c + di$ are complex numbers, then $(z\bar{z}) \cdot (w\bar{w}) = (zw) \cdot (\overline{zw})$.)

(b) Use (a) to show that whenever z can be written as the sum of two squares so can $z^5 + z$. Finally, use the fact that infinitely many integers (for example, all integers of the form $n^2 + 1$) can be written as the sum of two squares to complete the solution.

20th British Mathematical Olympiad, 1984

1. P, Q, and R are arbitrary points on the sides BC, CA, and AB respectively of triangle ABC. Prove that the triangle the vertices of which are the centres of the circles AQR, BRP, and CPQ is similar to triangle ABC.

(1) It may take a moment or two to realize that when the question refers to 'the circle AQR' it can only mean one thing—namely, the circle *through A, Q, and R*: that is, the *circumcircle* of the triangle AQR. Thus the *centre* of 'circle AQR' must be the *circumcentre* of triangle AQR (so you will need to know how to construct the circumcentre of a triangle). Now that you understand the question, you should draw an accurate diagram using ruler and compasses. (A good diagram can give clues about how to proceed—*and how not to proceed!* But for that it is important to make sure that ABC is a *general* triangle, and that the points P, Q, and R are positioned carefully to avoid 'accidents'.)

(2) One of the simplest of all problem-solving principles is *Name and Conquer*. To refer to something, it must have a name: and it often helps to choose a good name. You have to prove that the triangle the vertices of which are the circumcentres of AQR, BRP, and CPQ is 'similar to triangle ABC'. Part of your problem is to decide which circumcentre corresponds to A, which to B, and which to C. In the absence of other clues, it seems likely that A will correspond to the circumcentre nearest to A: so let A' denote the circumcentre of AQR, B' the circumcentre of BRP, and C' the circumcentre of CPQ. Your immediate task is to find some reason why the angles at A and at A' are equal (the reason will probably carry over to explain why the angles at B and B' are equal). Before proceeding, draw the three circles AQR (centre A'), BRP (centre B'), and CPQ (centre C') as accurately as you can.

(3) The first two circles meet at R and at a second point X (say). The second two circles meet at P and at a second point X' (say). The first and third circles meet at Q and at a second point X'' (say). Your diagram may lead you to suspect something about the three points X, X', and X''!

The quadrilateral $ARXQ$ is ****i*, so $\angle RXQ = 180° - \angle A$; similarly, $\angle RXP = 180° - \angle B$. Hence

$$\angle PXQ = 360° - (180° - \angle A) - (180° - \angle B) = \_\_\_\_.$$

It follows that quadrilateral $CPXQ$ is $****i*$, so $X' = X'' = X$. The rest is easy: $\angle A' = \angle B'A'X + \angle XA'C' = \frac{1}{2}\angle RA'X + \frac{1}{2}\angle XA'Q = \frac{1}{2}\angle RA'Q = \angle A.$

---

2. Let a_n be the number of binomial coefficients $\binom{n}{r}$ $(0 \leqslant r \leqslant n)$ which leave remainder 1 on division by 3, and let b_n be the number which leave remainder 2. Prove that $a_n > b_n$ for all positive integers n.

---

(1) The nth row of Pascal's triangle contains $n + 1$ numbers—namely $\binom{n}{0}, \binom{n}{1}, \dots, \binom{n}{n}$. Each leaves remainder 0, 1, or 2 on division by 3. If we are only interested in these remainders, it makes sense to replace each entry in Pascal's triangle by its residue (mod 3). Using the basic recurrence $\binom{n+1}{r} = \binom{n}{r} + \binom{n}{r-1}$ (mod 3), one can generate the triangle very quickly. Generate the next ten or eleven rows of the triangle on the right, keeping an eye open for any striking patterns which might give you a clue as to why the required result '$a_n > b_n$' should be true for all $n \geqslant 0$.

```
                    1                    n = 0
                                         n = 1
                 1     1                 n = 2
              1     2     1
           1     3     3     1
        1     4     6     4     1
     1     .     .     .     .     1
  .     .     .     .     .     .     .


                    1
                 1     1
              1     2     1
           1     0     0     1
        1     1     0     1     1
     1     2     1     1     2     1
  1     0     0     2     0     0     1
1     1     0     2     2     0     1     1
1  .     .     .     .     .     .     .     1
1     .     .     .     .     .     .     .     1
```

(2) The result '$a_n > b_n$' is certainly true for $n = 0, 1, 2$. One of the things that should have struck you in (1) is the way in which the next three rows $(n = 3, 4, 5)$ of Pascal's triangle (mod 3) seem to be built out of two copies of the first three rows with an inverted triangle of 0's in between. Since there are more 1s than 2s in rows 0, 1, 2, it follows that for each $n = 3, 4, 5$ the difference $a_n - b_n$ will be exactly twice the corresponding difference in row $n - 3$. The rows for $n = 6, 7, 8$ show that there is rather more to the problem than this! Nevertheless, it is important to formulate and prove the general fact underlying the above observation.

(a) Let $n = 3^k$. Prove that the coefficient of t^r in the expansion of $(1 + t)^n$ is divisible by 3 for all r, satisfying $1 \leqslant r \leqslant n - 1$. [Write $(1 + t)^{3^k} = ([1 + t]^3)^{3^{k-1}} = (1 + 3t + 3t^2 + t^3)^{3^{k-1}} = (1 + t^3)^{3^{k-1}} + \dots$. Then use induction on k.]

(b) Conclude that the rows $n = 3^k, 3^k + 1, \ldots, 3^k + 3^k - 1$ consist precisely of one copy of rows 0 to $(3^k - 1)$ on the extreme left and another on the extreme right, with an inverted triangle of 0s in between. Hence prove that $a_n > b_n$ whenever n has the form $3^k + i$ $(0 \leqslant i < 3^k)$ for some $k \geqslant 0$.

(3) It remains to prove that $a_n > b_n$ when n has the form $2 \cdot 3^k + i$ $(0 \leqslant i < 3^k)$. From what you have just proved, you know that the $(2 \cdot 3^k - 1)$th row consists of two end-to-end copies of '1 2 1 2 1 2 . . . 1 2 1'. Hence the $(2 \cdot 3^k)$th row has the form

'1 0 0 0 0 0 . . . 0 0 2 0 0 . . . 0 0 0 0 0 1'.

This then guarantees that the next 3^k rows will consist of one copy of the first 3^k rows on the extreme left, another copy of the first 3^k rows on the extreme right, and a third triangle in the middle (with its apex at the '2' in the centre of the $(2 \cdot 3^k)$th row), these three sub-triangles being separated by inverted triangles of zeros. The first three rows look like this:

```
    1 0 0 0 0 0 . . . . . 0 0 2 0 0 . . . . . 0 0 0 0 0 1
     1 1 0 0 0 0 0 . . . . 0 0 2 2 0 0 . . . . 0 0 0 0 0 1 1
    1 2 1 0 0 0 0 0 . . . 0 0 2 1 2 0 0 . . . 0 0 0 0 0 1 2 1
  . . . . . . . . . . . . . . . . . . . . . . . . . . . . . . . .
```

The central triangle contains lots of 2s, and at first sight it may not be clear how to prove that each row will still contain more 1s than 2s.

Fortunately, the central triangle is really very simple, for it is generated in exactly the same way as the left-hand and right-hand triangles—*except that it starts with '0, 2, 0' instead of '0, 1, 0' in the top row*. If we use the same Pascal's triangle recurrence, and simply multiply the initial row by some constant, then each subsequent row will be equal to the usual row multiplied by the same constant. Thus, whenever the (usual) triangle on the extreme left has a 1, the corresponding entry in the central triangle will be a ____, and whenever the (usual) triangle on the extreme left has a 2, the corresponding entry in the central triangle will be a ____. Thus the number of 1s in the left-hand triangle is equal to the number of 2s in the central triangle, and the number of 2s in the left-hand triangle is equal to the number of 1s in the central triangle. Hence these two triangles 'cancel each other out'—so the difference $a_n - b_n$ when $n = 2 \cdot 3^k + i$ $(0 \leqslant i < 3^k)$ is controlled by the corresponding difference in the triangle on the extreme right (namely $a_i - b_i$, which we already know is positive).

(4) The above approach stems from 'looking for obvious patterns' in Pascal's triangle (mod 3). It is in fact slightly easier to prove the result directly by induction *three rows at a time*. The result is certainly true when $n = 0, 1, 2$. Suppose that it is true for all rows up to row $3m - 1$. Use $(1 + t)^{3m} = (1 + t^3 + 3(t + t^2))^m$ to show that, if the mth row has the form '1, p, q, . . . , 1', then the $(3m)$th row has the form '1, 0, 0, p, 0, 0, q, 0, 0, . . . , 0, 0, 1', whence $a_{3m} > b_{3m}$. When you use this row to construct the $(3m + 1)$th row you find

that $a_{3m+1} =$ ____ and $b_{3m+1} =$ ____, so $a_{3m+1} > b_{3m+1}$. Constructing the
$(3m+2)$th row shows that $a_{3m+2} = 2a_m + b_m$ and $b_{3m+2} =$ ____, so $a_{3m+2} - b_{3m+2} = a_m -$ ____ > 0.
(5) Look at the actual values of $a_n - b_n$ (they are all *o*e** of **o). If you
think carefully about the above proofs, you should be able to find a formula
for $a_n - b_n$ (the exponent is related to the base 3 representation of n).

---

3. (a) Prove that, for all positive integers m,

$$\left(2 - \frac{1}{m}\right)\left(2 - \frac{3}{m}\right)\left(2 - \frac{5}{m}\right)\cdots\left(2 - \frac{2m-1}{m}\right) \leqslant m!.$$

(b) Prove that, if a, b, c, d, and e are positive real numbers, then

$$\left(\frac{a}{b}\right)^4 + \left(\frac{b}{c}\right)^4 + \left(\frac{c}{d}\right)^4 + \left(\frac{d}{e}\right)^4 + \left(\frac{e}{a}\right)^4 \geqslant \frac{b}{a} + \frac{c}{b} + \frac{d}{c} + \frac{e}{d} + \frac{a}{e}.$$

---

(1) Parts (a) and (b) look totally unrelated. Nevertheless, it makes sense to
start with part (a) (if only because the first part of such a question is generally
easier than the second). How many factors are there in the product $m!$ on the
RHS? And how many factors are there in the product on the LHS? It may
seem optimistic, but one reason why 'LHS $\leqslant$ RHS' might be true is if each
factor on the LHS happens to be less than or equal to the corresponding
factor on the RHS! (Note that $2 \times 8 \leqslant 3 \times 7$ even though $8 > 7$, so you have
no right to expect this to work—but it just might!)

(a) Prove that $2 - \frac{1}{m} \leqslant m$ and $2 - \frac{2m-1}{m} \leqslant 1$ for all $m \geqslant 1$.

(b) Is it true that $2 - \frac{2i+1}{m} \leqslant m - i$ for all i ($0 \leqslant i < m$)?

[Alternatively, reversing the order of the factors on the LHS and simplifying gives

$$\frac{1}{m} \cdot \frac{3}{m} \cdot \ldots \cdot \frac{(2m-1)}{m} \leqslant 1 \cdot 2^{m-1} \leqslant 1 \cdot 2 \cdot 3 \cdot \ldots \cdot m = m!]$$

(2) The methods outlined in (1) should have convinced you that the inequality in part (a) of the problem is rather weak. Show that

$$\left(2 - \frac{r}{m}\right)\left(2 - \frac{(2m-r)}{m}\right) \leqslant 1 \text{ for each } r.$$

Hence prove that the LHS of part (a) is $\leqslant 1$.

(3) Part (b) of the problem looks more challenging. If you knew a thing or two about inequalities, you would probably not be reading this in search of help; so we shall proceed slowly. The symmetry of the two sides should suggest a family of instances when the inequality becomes an equality (although it is less clear that this is the only way in which the two sides can be equal). However, this should convince you that the actual values of a, b, c, d, and e are irrelevant: it is the *$a*io*$ that matter. It therefore makes sense to simplify the notation slightly by writing $a/b = v$, $b/c = w$, $c/d = x$, $d/e = y$, and $e/a = z$, and to rewrite the inequality to be proved in terms of v, w, x, y, and z. Do this. (Note that you now have the extra piece of information that $vwxyz = \underline{}$.)

(4) Thus you have to prove an inequality involving *five* variables v, w, x, y, and z and *fourth* powers. One way of getting to grips with such a problem is to consider the corresponding problem for two and three variables first. This may prove nothing, but it can be a source of ideas:

(a) Write down the corresponding two-variable inequality (involving v and w only) and use the fact that $vw = \underline{}$ to show that the inequality is in fact an equality in this case.

(b) Write down the corresponding three-variable inequality in v, w, and x (involving *second* powers on the LHS, and with v, w, and x satisfying $vwx = 1$). Use $vwx = 1$ to simplify the RHS. At bottom every inequality comes down to proving that some expression is non-negative. In this case you have to prove that $v^2 + w^2 + x^2 - vw - wx - xv \geqslant 0$. The form of the terms on the LHS should suggest that you should try to write the LHS as a sum of perfect squares. (This is easiest if you first multiply both sides by 2.)

(5) Now return to the five-variable problem. You have to prove that

$$v^4 + w^4 + x^4 + y^4 + z^4 \geqslant \frac{1}{v} + \frac{1}{w} + \frac{1}{x} + \frac{1}{y} + \frac{1}{z}.$$

The first move is to use $vwxyz = 1$ to rewrite the RHS as $wxyz + \underline{} + \underline{} + \underline{} + \underline{}$. Now take these terms of the LHS, multiply the whole inequality by 2, and try to write the new LHS as a sum of perfect squares. (Since the problem with five variables is more complicated than that with three variables, you should not be surprised if you need an intermediate step: the obvious first move of writing five squares such as $(v^2 - w^2)^2$, introduces new terms ('$-2v^2w^2$', and so on). If you 'cancel' each such term (by adding '$2v^2w^2$', and so on), you obtain five new terms; these and the five old terms '$-(2wxyz + 2xyzw + 2yzvw + 2zvwx + 2vwxy)$' can then be rewritten (as in (4)) as the sum of five more perfect squares.)

[Alternatively, the fact that the LHS is a sum, together with the symmetry of the terms should suggest using the AM–GM inequality—although the fact that there are *i*e terms and that the exponent is *ou* argues strongly against a simple-minded application. The fourth powers suggest taking fourth roots; taking the terms on the LHS four at a time we obtain five different inequalities. Adding these inequalities gives the required result.]

---

4. Let N be a positive integer. Determine, with proof, the number of solutions x in the interval $1 \leqslant x \leqslant N$ of the equation $x^2 - [x^2] = (x - [x])^2$.

---

(1) Problems which involve an unfamiliar notation are often much easier than they seem. However, you may have to do some simple-minded calculations with small numbers just to get used to things.

(2) The variable x in the question can be any *real* number in the interval $1 \leqslant x \leqslant N$. The expressions $[x]$ and $[x^2]$ are easy to understand when x (and x^2) are i**e*e**, so it is natural to think about this case first:

(a) Suppose that x is an integer. What is the value of $x - [x]$? What is the value of $x^2 - [x^2]$? Does the equation $x^2 - [x^2] = (x - [x])^2$ hold in this case?

(b) Now suppose that x is not an integer, and let $[x] = n$ be the largest integer $\leqslant x$. Then $x = n + r$, where $0 < r < 1$. Calculate the value of the RHS $(x - [x])^2$ (in terms of r). Calculate x^2 (in terms of n and r). What value must $[x^2]$ take if the equation $x^2 - [x^2] = (x - [x])^2$ is to be satisfied?

(c) Suppose that $n = 1$; for how many values of r (with $0 < r < 1$) is $2nr$ an integer? Suppose that $n = 2$; for how many values of r (with $0 < r < 1$) is $2nr$ an integer? Suppose that $n = 3$; for how many values of r (with $0 < r < 1$) is $2nr$ an integer? In general, given any integer n, for how many values of r (with $0 < r < 1$) is $2nr$ an integer?

(d) You now know that solutions x of $x^2 - [x^2] = (x - [x])^2$, with $1 \leqslant x \leqslant N$, are of two kinds: integer solutions, of which there are exactly ____; and non-integer solutions of the form $n + r$ ($1 \leqslant n < N$, $0 < r < 1$), of which there are exactly $1 + 3 + 5 + \ldots +$ _____. Use this to obtain your answer in closed form.

---

5. A plane cuts a right circular cone with vertex V in an ellipse E, and meets the axis of the cone at C. The point A is an extremity of the major axis of E. Prove that the area of the curved surface of the slant cone with V as vertex and E as base is $(VA/AC) \times$ (area of E).

(1) No matter how you solve this problem you are going to have to struggle to picture things in three dimensions. So start by drawing as good a diagram as you can manage.

(2) One approach would be to try to calculate the curved surface area of the cone and the area of E directly. But this is not strictly necessary: all you have to do is to show that the *a*io of these two areas is equal to ____. Imagine a very thin (slightly curvy) triangle VPQ, with P and Q on the ellipse either side of, and equidistant from, the point A. Then PQ is *e**e**i*u*a* to VA, so the triangle VPQ has area $(PQ \cdot VA)/2$. Also, PQ is perpendicular to CA, so the triangle CPQ has area ____ . Thus the required ratio for the areas of the cone and the ellipse is in fact equal to the ratio of the areas of these two small triangles!

(3) Let $\angle CVA = \theta$ and $\angle VCA = \alpha$. Express VA/CA in terms of $\sin \theta$ and $\sin \alpha$. Let A' be the end of the major axis of E opposite A, and let P' and Q' be points equidistant either side of A' (with $P'Q'$ very small). Show that the ratio of the areas of the two triangles $VP'Q'$ and $CP'Q'$ is VA'/CA', and show that this is equal to VA/CA. (Observe that you were only told that 'A is an extremity of the major axis of E'; you were not told *which* extremity.)

(4) The previous two steps suggest an unexpected reason why the ratio of the two areas (of the cone and the ellipse) may be equal to VA/CA. Suppose you could show that for any two 'nearby' points P and Q on the ellipse, the ratio of the areas of VPQ and CPQ is exactly VA/CA. The result would then follow by decomposing the curved surface of the cone and the surface of the ellipse into a sequence of matching triangles. (This argument can be made rigorous using calculus.)

Let P and Q be two nearby points on the ellipse and let X be the midpoint of PQ. The calculation when X is neither at A nor at A' is a little more involved, since PQ is then perpendicular neither to VX nor to CX. Let A' be the lower of A and A'.

Cut the cone perpendicular to the axis through the point A' to create a circular base. By scaling, we may assume that the base has radius 1.

(a) Show that the angle between the plane of E and the horizontal is θ. When a region in this plane is projected onto the base plane, its area is multiplied by a fixed constant. What is that constant?

(b) Express the slant height VA' in terms of θ. Let the axis of the cone meet the base at the point C', and let the generators VP and VQ of the cone meet the base circle in P' and Q' respectively. If $\angle P'C'Q' = \gamma$, prove that $\angle PVQ = \gamma \cdot \sin \theta$. Conclude that the area of triangle VPQ is equal to $[VX^2 \cdot \gamma \sin \theta]/2$.

(c) Let $\angle VCX = \alpha'$. Calculate the length of the projection of CX on to the base plane. Hence show that the area of the projection of triangle CPQ on to the base plane is $(CX \sin \alpha')^2(\gamma/2)$. Now use the result of (a) to find the area of triangle CPQ.

(d) Hence show that area(VPQ)/area$(CPQ) = \sin \alpha/\sin \theta = VA/CA$, and so complete the solution.

> 6. Let a and m be positive integers. Prove that if there exists an integer x such that $a^2x - a$ is divisible by m, then there exists an integer y such that both $a^2y - a$ and $ay^2 - y$ are divisible by m.

---

(1) You are given positive integers a and m and an integer x such that $m \mid a(ax - 1)$. Note first that hcf$(a, ax - 1) = \_\_\_$. Thus m factorizes as $m = p \cdot q$, where $p \mid a$ and $q \mid (ax - 1)$ (and hcf$(p, q) = 1$).

You have to show that there exists an integer y such that both $y(ay - 1)$ and $a(ay - 1)$ are divisible by m. Now, since hcf$(m, a) = p$, $m \mid a(ay - 1)$ if and only if $q \mid (ay - 1)$—in which case $q = $ hcf$(m, ay - 1)$. Hence, m divides both $a(ay - 1)$ and $y(ay - 1)$ if and only if $p \mid y$ and $q \mid (ay - 1)$.

(2) You are told that $q \mid (ax - 1)$. Thus $q \mid (ay - 1)$ if and only if $q \mid [(ay - 1) - (ax - 1)] = \_\_\_\_$; that is, if and only if $q \mid (y - x)$. Thus you must prove that there exists a y of the form $kq + x$ which is also a multiple of p. To prove this, you must use the fact that hcf$(p, q) = \_\_$. Euclid's algorithm guarantees that there exist integers u and v such that $u \cdot p + v \cdot q = 1$, and hence $xu \cdot p + xv \cdot q = x$. Thus $y = xu \cdot p = (-xv) \cdot q + x$ has the required form and is visibly divisible by p as required. This completes the solution.

(3) A more direct approach (which produces a particular value of y) is to observe that $ay^2 - y = (a^2y - a)(y/a)$. Thus it makes sense to try to choose $y = za$ to be a multiple of a—since you can then forget one of the two messy conditions and concentrate on choosing y so that $m \mid (a^2y - a)$. That is, you must choose z such that '$m \mid (a^2x - a)$ (given) implies $m \mid (a^3z - a)$'. Show that $m \mid (a^2x - a)$ implies that $m \mid (a^3x^2 - a)$. Conclude that you may choose $z = \_\_\_$.

> 7. The quadrilateral $ABCD$ has an inscribed circle. To the side AB we associate the expression $u_{AB} = p_1(\sin \angle DAB) + p_2(\sin \angle ABC)$, where p_1 and p_2 are the lengths of the perpendiculars from A and B respectively to the opposite side CD. Define u_{BC}, u_{CD}, and u_{DA} similarly, using in each case the perpendiculars to the opposite side. Show that $u_{AB} = u_{BC} = u_{CD} = u_{DA}$.

---

(1) Let A denote $\angle DAB$, and so on. The definition of u_{AB} introduces the perpendicular distances p_1 and p_2, so it is natural to begin by expressing

each of these unfamiliar lengths in terms of something more familiar: for example, using the angle at D we have $\sin D = p_1/AD$, so $p_1 = AD \sin D$; similarly, $p_2 = BC \sin C$. Hence $u_{AB} = AD \sin D \sin A + BC \sin C \sin B$.

(2) You are told that the quadrilateral $ABCD$ 'has an inscribed circle'. Thus the side AD consists of a tangent from A and a tangent from D and the side BC consists of a tangent from B and a tangent from C. If t_A denotes the length of the tangent from the point A to the inscribed circle, and so on, then

$$(*) \qquad u_{AB} = (t_A + t_D)\sin D \sin A + (t_B + t_C)\sin C \sin B.$$

Since the RHS is unchanged when we swap A and D and swap B and C, $u_{AB} = u_{CD}$. Similarly, $u_{BC} = u_{AD}$.

(3) Thus it remains to prove that $u_{AB} = u_{AD}$. Since later questions tend to be more demanding than earlier questions, you should expect this to be harder. However, it is not all that hard. The expressions for u_{AB} and u_{AD} involve a curious mixture of lengths and trig functions. In (2) you managed to make partial progress despite this. Further exploration suggests one is unlikely to prove $u_{AB} = u_{AD}$ in the same way, so you must be prepared to use some trigonometry to rewrite u_{AB} and u_{AD} in the hope that they will become visibly equal!

(a) Let O be the centre of the inscribed circle. Use the fact that AO bisects $\angle BAD$ to express t_A in terms of r and $A/2$. Do the same for t_D, t_B, and t_C.

(b) Now substitute in the RHS of equation $(*)$ above and use the half-angle formula for 'sin' to prove that

$$u_{AB} = 4r\cos(A/2)\cos(D/2)(\sin(D/2)\cos(A/2) + \cos(D/2)\sin(A/2))$$

$$+ 4r\cos(B/2)\cos(C/2)(\sin(B/2)\cos(C/2) + \cos(B/2)\sin(C/2))$$

$$= 4r[\cos(A/2)\cos(D/2)\sin((A+D)/2)$$

$$+ \cos(B/2)\cos(C/2)\sin((B+C)/2)].$$

(c) Now use the fact that $(A+D)/2 = 180° - (B+C)/2$ to rewrite this as

$$u_{AB} = 4r[\cos(A/2)\cos(D/2)\sin((B+C)/2)$$

$$+ \cos(B/2)\cos(C/2)\sin((A+D)/2)].$$

Finally, expand $\sin((B+C)/2)$ and $\sin((A+D)/2)$ and rearrange to obtain an expression which is visibly equal to u_{AD}.

(4) Alternatively, you might like to show that $p_1 = r(1 + \cos D + \sin D \cot(A/2))$. Hence prove that $u_{AB} = r(\sin A + \sin B + \sin C + \sin D)$, which does not depend on 'AB'.

19th British Mathematical Olympiad, 1983

1. In triangle ABC with circumcentre O, $AB = AC$, D is the midpoint of AB, and E is the centroid of triangle ACD. Prove that OE is perpendicular to CD.

(1) There are three possible approaches here. If one could see a geometric reason *why* OE should be perpendicular to CD, then a purely geometric solution might turn out to be simplest. In the absence of an obvious geometric reason it is tempting to try using vectors, or even co-ordinate geometry:

(a) For a vector approach, let O be the origin, and let **a**, **b**, **c**, **d**, and **e** be the vectors for the points A, B, C, D, and E respectively. You should know a standard result which expresses **e** in terms of **a**, **c**, and **d**. You should also be able to write **d** in terms of **a** and **b**, and so express $\vec{CD}$ in terms of **a**, **b**, and **c**. Hence prove that the dot product $12\vec{OE} \cdot \vec{CD} = 3\mathbf{a} \cdot \mathbf{a} + \mathbf{b} \cdot \mathbf{b} - 4\mathbf{c} \cdot \mathbf{c} + 4\mathbf{a} \cdot \mathbf{b} - 4\mathbf{a} \cdot \mathbf{c}$. Finally, explain why $\mathbf{a} \cdot \mathbf{a} = \mathbf{b} \cdot \mathbf{b} = \mathbf{c} \cdot \mathbf{c}$, and use the fact that $\triangle ABC$ is isoceles to explain why $\mathbf{a} \cdot (\mathbf{b} - \mathbf{c}) = 0$. Hence complete the solution.

(b) For a co-ordinate approach you have to choose an origin. One good choice is to use the midpoint M of BC as origin, with MC as x-axis and MA as y-axis. Then $A = (a, 0)$, $C = (0, c)$, $B = (0, -c)$, $D = (\_\_, \_\_)$, and $E = (\_\_, \_\_)$. The circumcentre O lies on the y-axis, so has co-ordinates $(0, y)$ for some y. Now use $AO = CO$ to express y in terms of a and c: $y = \_\_\_\_$. Finally, calculate the gradients of OE and of CD to complete the solution.

(2) You should suspect that, if only you knew a little more, then you would understand that the result is no accident! Let F be the midpoint of AC and G be the centroid of $\triangle ABC$. Then G lies one third of the way up MA; in particular, D, G, and C are collinear. Moreover, DF is parallel to BC, and E lies on DF; hence GO is perpendicular to DE.

(a) Prove that $FE/FD = FG/FB$. Conclude that GE is parallel to BD—and hence perpendicular to DO.

(b) Hence show that O is the orthocentre of $\triangle DEG$. Why does this solve the problem?

2. The *Fibonacci sequence* $\{F_n\}$ is defined by

$$F_1 = 1, \quad F_2 = 1, \quad F_n = F_{n-1} + F_{n-2} \quad (n > 2).$$

Prove that there are unique integers a, b, and m such that $0 < a < m$, $0 < b < m$, and $F_n - anb^n$ is divisible by m, for all positive integers n.

(1) Any problem with four parameters (n, a, b, and m) and a sequence (F_n) can be a little off-putting at first. It is hard to know how to begin. However, if '$F_n - anb^n$' is to be divisible by m for *all* positive integers n', then it must certainly work when $n = 1$, 2, or 3. So why not start by investigating what this tells you about a, b, and m?

(a) m divides $F_1 - a \cdot 1 \cdot b$ if and only if $m \mid ab - 1$. Use this and the fact that m must divide $F_2 - a \cdot 2 \cdot b^2$ to prove that $m \mid 2b - 1$.

(b) Use (a) and the fact that m must divide $F_3 - a \cdot 3 \cdot b^3$ to prove that $m \mid b^2 + b - 2$. Factorize $b^2 + b - 2$, and use the fact that $m \mid 2b - 1$, to conclude that $m \mid b + 2$. Hence m must divide $2(b + 2) - (2b - 1) =$ ____ .

(c) Since we are asked to find integers m (and a) satisfying $m > a > 0$, this leaves only one possible value for m—namely $m =$ ___ . Since $m > b > 0$ and $m \mid b + 2$, there is only possible value of b—namely $b =$ ___ . And since $m \mid ab - 1$, there is only one possible value of a—namely $a =$ ___ .

(d) It remains to prove (by induction, using the given recurrence) that $F_n \equiv 2n \cdot 3^n \pmod{5}$ for all $n \geqslant 1$.

---

3. Given any real number $a \neq -1$, the sequence $x_1, x_2, x_3, \ldots$ is defined by

$$x_1 = a, \quad \text{and} \quad x_{n+1} = x_n^2 + x_n \quad \text{for all } n \geqslant 1.$$

Let $y_n = 1/(1 + x_n)$. Let S_n be the sum, and let P_n be the product, of the first n terms of the sequence $y_1, y_2, y_3, \ldots$. Prove that $aS_n + P_n = 1$, for all n.

---

(1) Like the previous question, this one introduces so much notation that it is hard to tell whether it is really complicated, or just something simple wrapped up to *look* complicated.

(a) Check that $S_1 = 1/(a + 1) = P_1$. Hence verify that $aS_1 + P_1 = 1$.

(b) Suppose that $aS_n + P_n = 1$ for some $n \geqslant 1$. Show that this would imply that $aS_{n+1} + P_{n+1} = 1$ *provided that* $P_n = a/(x_{n+1})$.

(c) Check that $P_1 = a/x_2$. Prove (by induction on n) that $P_n = a/x_{n+1}$ for all $n \geqslant 1$. Hence solve the problem.

4. The two cylindrical surfaces

$$x^2 + z^2 = a^2, \qquad z > 0, \quad |y| \leqslant a,$$

and

$$y^2 + z^2 = a^2, \qquad z > 0, \quad |x| \leqslant a,$$

intersect. Together with the plane $z = 0$ they enclose a dome-like shape which we shall call a *cupola*. The cupola is placed on top of a vertical tower of height h, the horizontal cross-section of which is a square of side $2a$. Find the shortest distance over the surface of the cupola and tower from the highest point of the cupola to a corner of the base of the tower.

(1) The main difficulty in solving three-dimensional problems is often the first step of 'seeing' exactly what is going on. So draw the best diagram you can, to get a feeling for the solid which is the intersection of these two (half-)cylinders.

(2) Each (half-)cylinder has diameter $2a$ and length $2a$, so the base of their intersection is a square of side ____ (just the right size to fit on top of the vertical tower). The uppermost point of their intersection is at the point at which the horizontal upper 'generators' of the two cylinders cross. Four (curved) lines run from this highest point to the four corners of the square base: these are the lines where the two cylinders cut into each other. Hence these curved lines lie on the surface of both cylinders; the rest of the surface of the intersection of the two cylinders lies inside one of the two cylinders. Thus the surface of the solid of intersection consists of four 'curvy-triangular' sections—each being part of the surface of one of the two cylinders. Unlike a sphere, the surface of a cylinder can be laid out flat. Thus each of the cupola's four curvy triangles can be 'unrolled' so that it is a direct continuation of the vertical wall of the tower to which it is joined. The shortest distance from the apex A to the corner C of the base is therefore given by drawing the straight line AC joining the two points on this 'flattened' picture. The flattened curvy triangle has height ____ (since its altitude was originally one quarter of a circle of radius a), so the 'tower plus curvey triangle' has height $AM = h + \pi a / 2$. Now use $AC^2 = AM^2 + MC^2$ to calculate AC.

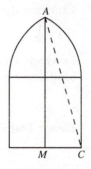

5. Given ten points inside a circle of diameter 5, prove that the distance between some pair of the given points must be less than 2.

(1) This may look tantilisingly familiar, yet different! If you have ever seen anything like it before, you will expect to have to use the *pigeonhole principle*: if $N + 1$ points are arranged in N boxes, then some box must contain at least two points. In this case you are told that there are *ten* points in a circle of diameter 5: so if you cut up the circle into *nine* parts, some part must contain at least two points. Unfortunately, this will only solve the problem if you choose the parts so that they all have 'diameter' < 2 (since then the distance between the two points which end up in the same part will be less than 2).
(2) Unfortunately, the most obvious ways of cutting a circle of diameter 5 into nine pieces don't work. This doesn't mean that the idea is useless—only that you must use it more imaginatively. The 'obvious' ways of cutting a circle into nine pieces tend to assume that all the pieces are more or less the same shape. For example, if you treat the circle as a cake and cut it into nine equal 'slices', you face the problem that each slice has 'diameter' $\frac{5}{2}$ (equal to the radius of the circle), and $\frac{5}{2}$ (> 2) is too large.
(3) You need an idea. Suppose that you start by cutting out a circle of radius $r = \_\_$ from the centre of the large circle. If two points were to end up *inside* this small circle, the distance between them would certainly be less than $2r$ (so you want $2r \leqslant 2$). You now need to cut up the outer ring (between this small circle and the large circle) into *eight* pieces, and to check that each of these pieces has 'diameter' less than 2. Checking the diameter of each piece is easiest if all the pieces have exactly the same shape; so if you want an easy life, there is really only one way to cut up the outer ring!

---

6. Consider the equation

   (*) $\sqrt{(2p + 1 - x^2)} + \sqrt{(3x + p + 4)} = \sqrt{(x^2 + 9x + 3p + 9)}$

   in which x and p are real numbers and the square roots are to be real (and non-negative). Show that if x and p satisfy (*), then

   $$(x^2 + x - p)(x^2 + 8x + 2p + 9) = 0.$$

   Hence find all real numbers p for which (*) has just one solution x.

---

(1) It is always worth hesitating briefly before charging ahead and squaring both sides of an equation of the form $\sqrt{a} + \sqrt{b} = \sqrt{c}$, since this can sometimes lead one into an algebraic quagmire. However, in this instance you should be able to see that the RHS of the resulting equation

(**) $2(\sqrt{a} \cdot \sqrt{b}) = c - b - a$

simplifies rather nicely.

(a) Square both sides of the given equation and rearrange it into the form (**).

(b) Now square both sides of (**), take all the terms to one side and factorize into the required form.

(2) You now know that, if x satisfies the original equation, then one of the two factors $(x^2 + x - p)$ or $(x^2 + 8x + 2p + 9)$ must be zero: for each given value of p this gives *ou* possible values of x. Unfortunately, some of these pairs p, x do not solve the original equation, since squaring introduces spurious solutions. However, you have no real choice but to suppose that p is given, and to examine the four possible values of x in detail.

(a) Suppose that p is given. Write down the two roots x_1 and x_2 of $x^2 + x - p = 0$, and the two roots x_3 and x_4 of $x^2 + 8x + 2p + 9 = 0$.

(b) If x satisfies $x^2 + 8x + 2p + 9 = 0$, then $x^2 = -(8x + 2p + 9)$. Use this to show that $2p + 1 - x^2 = -2(x + 2)^2 \leqslant 0$, and that $3x + p + 4 = -(\frac{1}{2})(x + 1)^2 \leqslant 0$, and that $x^2 + 9x + 3p + 9 = (2x + 3)^2 \geqslant 0$. Hence show that, no matter what the value of p may be, x_3 and x_4 can never be solutions of the original equation.

(c) Let x_1 be the larger of x_1 and x_2. If x satisfies $x^2 + x - p = 0$, then x is real if and only if $p \geqslant \underline{\quad}$. Moreover, $p = x^2 + x$; use this to show that $2p + 1 - x^2 = (x + 1)^2 \geqslant 0$, that $3x + p + 4 = (x + 2)^2 \geqslant 0$, and that $x^2 + 9x + 3p + 9 = (2x + 3)^2 \geqslant 0$. Hence, given any value of $p \geqslant -(\frac{1}{4})$, x_1 is a solution of the original equation *provided that* $|x_1 + 1| + |x_1 + 2| = |2x_1 + 3|$, and x_2 is a solution of the original equation *provided that* $|x_2 + 1| + |x_2 + 2| = |2x_2 + 3|$.

(d) Show that $|x_1 + 1| + |x_1 + 2| = |2x_1 + 3|$ for each $p \geqslant -(\frac{1}{4})$.

(e) Show that $|x_2 + 1| + |x_2 + 2| = |2x_2 + 3|$ provided that $p \geqslant 2$, or $-(\frac{1}{4}) \leqslant p \leqslant 0$.

(f) Conclude that, if p is given, then the original equation has a unique solution x if and only if either $x = x_1 = x_2$ and $p = -(\frac{1}{4})$, or $x = x_1$ and $0 < p < 2$.

18th British Mathematical Olympiad, 1982

1. The convex quadrilateral $PQRS$ has area A; O is a point inside $PQRS$. Prove that if

$$2A = OP^2 + OQ^2 + OR^2 + OS^2,$$

then $PQRS$ is a square with O as its centre.

(1) Faced with an unusual formula like this one, it is all too easy to lose sight of the information that is given and to go round and round in circles. The RHS could arise in a number of ways (from Pythagoras, from the cosine rule, or even from squaring algebraic expressions involving OP, OQ, and so on). You are given two major clues.

(a) First, you are told that the RHS is equal to '$2A$'. Thus you should perhaps begin by finding a suitable expression for the area of the quadrilateral that involves the four lengths OP, OQ, OR, and OS. Do this (there is only one real choice).

(b) Second, you are told to *prove* that the given formula implies that the quadrilateral must be a square *with O at its centre*. Thus, part of your conclusion must be that $OP = OQ = OR = OS$; that is, that the differences $OP - OQ$, $OQ - OR$, $OR - OS$, and $OS - OP$ are all equal to ____. Combined with the form of the RHS, this suggests that it might be enough to prove that

$$(*) \quad (OP - OQ)^2 + (OQ - OR)^2 + (OR - OS)^2 + (OS - OP)^2 = 0.$$

(2) You know that the LHS of $(*)$ is $\geqslant 0$. You also know from the formula in the question that, when you expand the terms on the LHS of $(*)$, the sum of the squared terms $2(OP^2 + OQ^2 + OR^2 + OS^2)$ is equal to ____. Now combine the expression for A from (1)(a), and the fact that $0 \leqslant \sin\theta \leqslant 1$ whenever $0 < \theta < 180°$, to deduce that the sum of the mixed terms $2OP \cdot OQ + 2OQ \cdot OR + 2OR \cdot OS + 2OS \cdot OP$ in $(*)$ is $\geqslant$ ____. Conclude that the LHS of $(*)$ is also $\leqslant 0$. Use this to complete the solution.

---

2. When written in base 2, a multiple of 17 contains exactly three digits 1. Prove that it contains at least six digits 0, and that if it contains exactly seven digits 0, then it is even.

---

(1) Let N be such an integer. Then $N = 2^a + 2^b + 2^c$, with $a > b > c$; N is divisible by 17 if and only if $N' = N/(2^c) = 2^{a'} + 2^{b'} + 1$ is divisible by 17 (where $a' = a - c$, $b' = b - c$).

(a) How do powers of 2 behave (mod 17)? Write out the sequence of remainders (mod 17) for $2^0 = 1, 2^1, 2^2, \ldots$. Explain why 2^{8+i} has the same remainder as 2^i.

(b) Show that the only way in which $2^{a'} + 2^{b'} + 1$ (with $a' > b' > 0$) can be a multiple of 17 is if the three remainders '$2^{a'}$, $2^{b'}$, 1 (mod 17)' are (i) '8, 8, 1' or (ii) '1, −2, 1' (or '−2, 1, 1'). In the first case, $a' = 8k + 3$ and $b' = 8m + 3$ (with $k > m \geqslant 0$), so N' has at least 12 digits, and hence at least nine zeros. In the second case one of a' and b' has the form $8k$ with $k \geqslant 1$, so N' has at least *i* digits, and hence at least *i* zeros.

(c) Conclude that if $N = 2^a + 2^b + 2^c$ has exactly seven digits zero, then N' cannot be of the form (b)(i). Thus N must be even, with $N' = N/2$ of the form (b)(ii).

(2) Alternatively, you may observe that $17 = 10001$ (*base 2*). Let $N = 17m$. If $m < 16$, then $m = wxyz$ (*base 2*) must have four or fewer binary digits; but then $17m = wxyzwxyz$ (*base 2*) would have an *even* number of 1's. Thus $m \geqslant 16$, so $17m \geqslant$ ____ has at least $*i*e$ digits—exactly three of which are 1's; hence $17m$ has at least $*i*$ zeros. This idea can be sharpened to complete the second part.

---

3. If $s_n = 1 + \dfrac{1}{2} + \dfrac{1}{3} + \dfrac{1}{4} + \ldots + \dfrac{1}{n}$ and $n > 2$, prove that

$$n(n+1)^a - n < s_n < n - (n-1)n^b,$$

where a and b are given in terms of n by $an = 1$ and $b(n-1) = -1$.

---

(1) Some problems are excellent Olympiad problems because there are many possible approaches, all of which require the solver to demonstrate ingenuity. Other problems challenge the solver to find the one approach that is likely to work. This problem is of the latter kind—although you should not need divine inspiration to find the right method. The problem involves inequalities, and the key ingredient s_n is, by definition, a $*u*$. If you consider the first inequality to be proved, namely

$$n(n+1)^a - n < s_n,$$

the main term on the LHS is a $**o*u**$. Moreover, the product on the LHS involves nth roots (since $a = 1/n$), and you have to prove that this product is less than the sum $s_n + n$. All of this should suggest strongly the need to use the ____ – ____ inequality.

(2) The fact that the term $(n + 1)^a$ involves nth roots suggests further that the sum $s_n + n$ should be rewritten as a sum of n (rather than $n + 1$) terms; the only obvious way of doing this is to share out the extra term n equally between the n terms of the form $1/r$, so that each becomes $(1/r) + 1 =$ ____. And the idea of using the AM–GM inequality suggests that the factor n in the product on the LHS should be taken to be the other side so that $(s_n + n)/n$ becomes the arithmetic mean of these n terms.

Everything now falls miraculously into place. The n terms in the sum are all different, so the AM–GM inequality becomes a strict inequality; and the n rewritten terms $(1 + r)/r$ in the sum are such that their product cancels beautifully, giving the required inequality.

(3) Now do something similar for the sum '$(n - s_n)/(n - 1)$' and for the product n^b to prove the second inequality.

---

4. For each choice of real number u_1, a sequence $u_1, u_2, u_3, \ldots$ is defined by the recurrence relation $u_n^3 = u_{n-1} + \frac{15}{64}$ $(n \geqslant 2)$. By considering the curve $x^3 = y + \frac{15}{64}$, or otherwise, describe, with proof, the behaviour of u_n as n tends to infinity.

---

(1) You have to 'describe the behaviour of the sequence $\{u_n\}$ as n tends to infinity'. 'Behaviour' here presumably means: Does the sequence tend to a (finite) limit? Does it diverge to $+\infty$ or to $-\infty$? Or does it wobble about? In general, one expects the answer to depend on the starting value u_1. For example, the sequence defined by the recurrence relation $x_n = 2/x_{n-1}$ converges to $\sqrt{2}$ if $u_1 > 0$, and converges to $-\sqrt{2}$ if $u_1 < 0$.

(2) In this problem the recurrence relation $u_n^3 = u_{n-1} + \frac{15}{64}$ is such that it is difficult to investigate exactly what happens for particular starting values (no calculators!). One way to get started is to suppose that (for some choice of u_1) the sequence $u_1, u_2, u_3, \ldots$ tends to a limit, x say, and to try to calculate the possible values of x.

(a) If the sequence converges to x, then, as n increases, both u_n and u_{n-1} tend to x. Thus x satisfies the equation $x^3 = x + \frac{15}{64}$. Use the Remainder Theorem to find one root α of this equation. Hence factorize the expression $f(x) = x^3 - x - \frac{15}{64}$ as a product of one linear factor and one quadratic factor. Then find the two other roots β and β' ($\beta < 0 < \beta'$). Hence there are at most three different limits for the sequence $\{u_n\}$.

(b) Calculate the first four values u_1, u_2, u_3, and u_4 of the sequence:

   (i) when $u_1 = \alpha$;   (ii) when $u_1 = \beta$;   (iii) when $u_1 = \beta'$.

   Conclude that the limits α, β, and β' can all occur, so there are *exactly* three different possible limits depending on the starting value u_1.

(c) Try to obtain a rough idea of how the sequence $u_1, u_2, u_3, \ldots$ behaves (i) when $u_1 = -2$, (ii) when $u_1 = -1$, (iii) when $u_1 = 0$, (iv) when $u_1 = 1$, and (v) when $u_1 = 2$.

(3) You should now be in a position to use 'the curve $x^3 = y + \frac{15}{64}$' to give a complete analysis of how the sequence behaves.

(a) Make an accurate sketch of the curve $x^3 = y + \frac{15}{64}$. On the same pair of axes, draw the line $y = x$.

(b) Choose an initial value u_1 for y and draw the horizontal line $y = u_1$ until it cuts the curve (at $x = u_2$). Then draw the vertical line $x = u_2$ until it cuts the line $y = x$ (at the point (____, ____)). Then draw the horizontal line $y = u_2$ until it cuts the curve (at $x =$ ____). Continue in this fashion. (This procedure is very like the Newton–Raphson procedure for finding approximate roots of an equation.)

(c) What happens if your initial value u_1 is less than β? What happens if $\beta < u_1 < \alpha$? What happens if $\alpha < u_1 < \beta'$? What happens if $\beta' < u_1$?

(4) You should now think that you know the answer to the problem. So all that remains is to find some way of presenting the *proof*.

(a) Suppose that $u_1 < \beta$. Prove: (i) $u_n < \beta$, for all $n \geq 1$; (ii) $u_{n-1} < u_n$ for all $n > 1$. Thus $\{u_n\}$ is an increasing sequence, bounded above by β, and so must converge to some real number $\leq \beta$. (To conclude that the sequence converges to β, you must appeal to what you proved in (2)(a); namely, that there are only three possible limiting values for $\{u_n\}$, and that only one of these is $\leq \beta$.)

(b) Suppose that $\beta < u_1 < \alpha$. Prove: (i) $\beta < u_n < \alpha$, for all $n \geq 1$; (ii) $u_{n-1} < u_{\ldots}$ for all $n > 1$. Conclude that $\{u_n\}$ converges to α.

(c) Now do something similar when $\alpha < u_1 < \beta'$, and when $\beta' < u_1$.

---

5. A right circular cone stands on a horizontal base of radius r. Its vertex V is at a distance l from each point on the perimeter of the base. A plane section of the cone is an ellipse with lowest point L and highest point H. On the curved surface of the cone, to one side of the plane VLH, two routes, R_1 and R_2, from L to H are marked: R_1 follows the semi-perimeter of the ellipse, while R_2 is the route of shortest length. Find the condition that R_1 and R_2 intersect between L and H.

---

(1) Draw a decent diagram of the cone and the ellipse, marking the points V, L, and H. Draw another diagram of the cone cut along VH and laid flat, marking the points V, L, and H and the line VL. Mark the shortest route R_2 ($= HL$) on the 'flattened' cone.

The particular diagrams that you have drawn may tempt you to jump to the conclusion that the two paths R_1 and R_2 can never intersect! However, diagrams can mislead. There are a number of parameters here: the inclination of the plane can vary—thereby changing the relative positions of H and L; and the angle of the flattened circular sector—after cutting the cone along VH—can vary from almost 0° to nearly 360°.

(2) What angle does the semi-perimeter R_1 of the ellipse make at the point H with the generator VH? What angle does R_1 make at L with the generator VL?

Suppose that the plane section is perpendicular to the axis of the cone.

Then all points of R_1 are equidistant from V. Hence, in the diagram of the 'flattened' cone, R_1 is the circular arc LH centred at V, so the paths R_1 and R_2 do not intersect between L and H.

Thus you may assume that the plane is genuinely tilted and that L is below H. Let X be an arbitrary point on the semi-perimeter R_1 of the ellipse. As X moves from H to L along R_1, what happens to the length VX?

(3) On the flattened cone, the ellipse R_1 sets out from L perpendicular to the line VL; that is, R_1 starts out from L on the opposite side of LH ($= R_2$) from V.

(a) Suppose that R_1 also sets out from H on the opposite side of LH from V. Then the two paths R_1 and R_2 must cross an e*e* number of times. A complete solution should prove that, in this case, the two paths do not cross at all. However, this is not obvious and depends on a tricky calculation! (Let O be the centre of the base; let VL, VH, and VX meet the base of the cone, at L', H', and X'. Take OV and OH' as the z- and y-axes, and let C be the point on the circular base such that OC is the x-axis. Thus the co-ordinates of X' in terms of $\angle COX' = \phi$ are $(r \cos \phi, r \sin \phi, 0)$. Show that the co-ordinates of the point X (in terms of ϕ and the distance ρ_x of X from the z-axis OV) are $(\rho \cos \phi, \rho \sin \phi, OV - \rho \cot \alpha)$, where $\sin \alpha = r/l$. Write $VX = R$ and find the polar equation of R_1 in the plane of the flattened out cone. Hence show that R_1 can have at most one point of inflexion, and so cannot cross LH more than once.)

(b) Suppose next that R_1 sets out from H *on the same side of LH as V*; that is, suppose that in the flattened cone, $\angle VHL > 90°$. Then R_1 must at some point cross the line segment R_2. In the flattened cone show that $\angle HVL = \pi r/l$; hence rewrite the condition $\angle VHL > 90°$ as an inequality $VH < VL \cos(\pi r/l)$.

---

6. Prove that the number of sequences $a_1, a_2, \ldots, a_n$, with each $a_i = 0$ or 1 and containing exactly m occurrences of '01', is $\dbinom{n+1}{2m+1}$.

---

(1) Combinatorics problems can often be solved very simply, but the easy solution almost always comes after one has struggled to find a simple-minded but messy solution!

The most simple-minded approach here is to count the number of sequences for each possible position of the *first* occurrence of '01', and then add the results.

(a) Let $f(n; m)$ denote the number of possible sequences of length n with exactly m occurrences of '01'. If the first '01' occurs in the first two positions, then there must be exactly ____ occurrences of '01' in the remaining ____ positions, so there are exactly $f(n-2; m-1)$ possible sequences of this type. (Strictly speaking, to prepare for the general case, you should say that there are exactly $f(0;0)$ ways of filling the first 0 positions and $f(n-2; m-1)$ ways of filling in the last $n-2$ positions, and hence $f(0;0)f(n-2; m-1)$ ways altogether—where one clearly needs the convention $f(0;0)=1$.)

If the *first* occurrence of '01' is in the second and third positions, then there are exactly $f(1;0)$ ways of filling the first position, and $f(\_\_\_;\_\_\_)$ ways of filling the last $n-3$ positions, and hence $f(1;0)f(\_\_\_;\_\_\_)$ ways altogether.

(b) Repeat this for the general case (in which the first occurrence of '01' is in the ith and $(i+1)$th positions). Hence derive the recurrence

$$f(n;m) = \sum_{i=0}^{n-2} f(i;0)f(n-i-2; m-1).$$

(c) Finally, check that $f(n;0) = n+1 = \binom{n+1}{1}$ for every n and then use induction to change the RHS into a familiar expression involving binomial coefficients (see 1985, Question 3).

(2) A variation on (1) is to partition the collection of all possible sequences into just two types: (a) those that end with 0, and (b) those that end with 1. Let $A(n; m)$ be the number of sequences of type (a) which have length n with m 01's, and let $B(n; m)$ be the number of sequences of type (b). Thus

$$f(n;m) = A(n;m) + B(n;m).$$

(i) Explain why $A(n; m) = f(n-1; m)$.

(ii) Show that $B(n; m) = f(n-2; m-1) + B(n-1; m)$.

(iii) Guess an expression for $B(n; m) = $ ____ (as a binomial coefficient), and prove both $B(n; m) = $ ____ and $f(n;m) = \binom{n+1}{2m+1}$ simultaneously by induction.

(3) An alternative approach is to reformulate the question in some way to explain the simple form of the answer $\binom{n+1}{2m+1}$. Each sequence $a_1, a_2, \ldots, a_n$ has not only n 'positions', but also $n+1$ 'gaps' between terms—including the two ends. Thus it looks as though you have to explain why the number of sequences with exactly m 01's is the same as the number of ways of choosing $2m+1$ of these 'gaps'.

Once you have had this idea, it should not take long to realize that each occurrence of '01' involves a jump *up* (from 0 to 1), and that before the next occurrence of '01' the terms of the sequence must jump back *down* (from 1 to 0). Thus each sequence with exactly m occurrences of '01' is uniquely determined by specifying the $2m + 1$ gaps (chosen from the $n + 1$ possible gaps) at which the sequence 'jumps'—with the sequence remaining constant between successive selected gaps: there are exactly $\binom{n+1}{2m+1}$ such sequences —provided that one specifies that all sequences are deemed to start with a string of 0's unless the very first 'gap' (to the left of the first term) is among those chosen, in which case the sequence jumps from 0 to 1 before it starts, and so starts with a string of 1's.

17th British Mathematical Olympiad, 1981

1. The point H' is the orthocentre of triangle ABC. The midpoints of BC, CA, and AB are A', B', and C' respectively. A circle with centre H cuts the sides of triangle $A'B'C'$ (produced if necessary) in six points: D_1 and D_2 on $B'C'$; E_1 and E_2 on $C'A'$; and F_1 and F_2 on $A'B'$. Prove that $AD_1 = AD_2 = BE_1 = BE_2 = CF_1 = CF_2$.

(1) This question involves so many points and lines that it is crucial to draw a decent diagram—using ruler and compasses (and a sharp pencil).

(2) Half of what you have to prove is straightforward. Let AH meet $B'C'$ at the point L. Since H is the centre of the circle, $HD_1 = HD_2$ and $LD_1 = LD_2$. Moreover, the line $D_1D_2 = B'C'$ is *a*a**e* to BC, and hence *e**e**i*u*a* to AH. Thus $AD_1 = AD_2$ (Why?). Similarly, $BE_1 = BE_2$ and $CF_1 = CF_2$.

(3) It remains to show that $AD_1 = BE_1$. This is not obvious, and requires a calculation (using Pythagoras, or vectors).

(a) Let BH meet $C'A'$ at M. Then $AL^2 + LC'^2 = AC'^2 = BC'^2 = BM^2 + MC'^2$ (since C' is the *i**oi** of AB).

(b) $HL^2 + LD_1^2 = HD_1^2 = HE_1^2 = HM^2 + ME_1^2$ (since H is the *e***e of the circle through D_1 and E_1).

(c) $HL^2 + LC'^2 = HC'^2 = HM^2 + MC'^2$ (since $\angle HMC' = \angle HLC' = 90°$).

(d) Adding (a) and (b) and subtracting (c) gives

$$(AL^2 + LC'^2) + (HL^2 + LD_1^2) - (HL^2 + LC'^2)$$
$$= (BM^2 + MC'^2) + (HM^2 + ME_1^2) - (HM^2 + MC'^2).$$
$$\therefore AD_1^2 = AL^2 + LD_1^2 = BM^2 + ME_1^2 = BE_1^2.$$

A similar calculation shows that $BE_1 = CF_1$.

(4) Alternatively, let H be the origin of (vector) co-ordinates, and let $\mathbf{a}$, $\mathbf{b}$, and $\mathbf{c}$ be the vectors for points A, B, and C. Then $\mathbf{a} \cdot (\mathbf{b} - \mathbf{c}) = 0$ (since AH is ∗e∗∗e∗∗i∗u∗a∗ to BC); similarly, $\mathbf{b} \cdot (\mathbf{c} - \mathbf{a}) = 0$, so $\mathbf{a} \cdot \mathbf{b} = \mathbf{b} \cdot \mathbf{c} = \mathbf{c} \cdot \mathbf{a} = k$ (say). Now let $\mathbf{a}'$, $\mathbf{b}'$, and $\mathbf{c}'$ be the vectors for A', B', and C'. Then $\mathbf{b}' = \underline{\quad}$ and $\mathbf{c}' = \underline{\quad}$. Since D_1 lies on $B'C'$, its vector $\mathbf{d}_1 = t\mathbf{b}' + (1 - t)\mathbf{c}'$. Moreover, D_1 lies on the circle centre H—of radius r (say); thus $\mathbf{d}_1 \cdot \mathbf{d}_1 = r^2$. Now show that $|AD_1|^2 = (\mathbf{d}_1 - \mathbf{a}) \cdot (\mathbf{d}_1 - \mathbf{a}) = r^2 - k$. Since this depends only on r and k, the result follows.

---

2. Given positive integers m and n, S_m is equal to the sum of m terms of the series

$$(n + 1) - (n + 1)(n + 3) + (n + 1)(n + 2)(n + 4)$$

$$- (n + 1)(n + 2)(n + 3)(n + 5) + \ldots,$$

the terms of which alternate in sign, with each term (after the first) equal to the product of consecutive integers with the last but one integer omitted. Prove that S_m is divisible by $m!$, but not necessarily by $m!(n + 1)$.

---

(1) This looks horrible, but it is only a simple question dressed up to look difficult. Throughout, we assume that n is fixed.

(a) Clearly, $S_1 = n + 1$. Calculate S_2 (in fully simplified—that is, factorized—form). Show that $S_2/2!$ is odd when n is odd, so that S_2 is not divisible by $2!(n + 1)$ when n is odd. (This observation answers the second part of the question.)

(b) Now calculate S_3 (in fully simplified form).

(c) State what you expect the simplified form of S_4 to be. Then do the necessary calculation to check that your guess is correct.

(d) You now want to prove that $S_m = (-1)^{m+1}(n + 1)(n + 2) \cdots (n + m)$. This is clearly true when $m = 1$. And if it is true for some $m \geqslant 1$, then

$$S_{m+1} = S_m + (-1)^{m+2}(n + 1)(n + 2) \cdots (n + m)(n + m + 2)$$

$$= (-1)^{m+2}(n + 1)(n + 2) \cdots (n + m)[-1 + (n + m + 2)]$$

$$= \underline{\hspace{6cm}}.$$

---

3. Let a, b, and c be any positive numbers. Prove that:
   (a) $a^3 + b^3 + c^3 \geqslant a^2b + b^2c + c^2a$;
   (b) $abc \geqslant (a + b - c)(b + c - a)(c + a - b)$.

(1) Harmless looking inequalities like the one in part (a) can be maddeningly elusive: there are always so many more ways to fail than to succeed. The simplest principle for inequalities depends on the fact that squares are *non-negative*; yet here we have *cubes* (see (3) below). The LHS suggests using the AM–GM inequality; but the form of the RHS means that one cannot use this in the 'obvious' way (see (4) below).

(2) You may use the cyclic symmetry of the two sides to assume (without loss of generality) that c (say) is the smallest of the three variables. However, it is not at all clear that you can also assume that $a \geqslant b$ (or $b \geqslant a$!).

One thing that you should certainly notice, and bear in mind all the time, is that the inequality becomes an equality when $a = \underline{\hspace{1cm}} = \underline{\hspace{1cm}}$. Another is the fact that the RHS and LHS are both *homogeneous* of the same degree (that is, all terms are of the same degree); so you are free to scale the variables to make one of them—c say—equal to 1.

(i) Put $c = 1$, $b = 1 + y$, and $a = 1 + x$ $(x, y \geqslant 0)$. Then expand the two sides and collect up terms to show that the inequality to be proved is equivalent to proving '$(x - y)^2 + x^3 + y^3 + x^2 + y^2 - x^2 y \geqslant 0$'.

(ii) Observe that, if $x \geqslant y$, then $x^3 - x^2 y \geqslant 0$, and if $y \geqslant x$, then $y^3 - x^2 y \geqslant 0$. Hence complete the proof of (a).

(3) Alternatively, since you know that a, b, and c are *o*i*i*e, the fact that 'squares are $\geqslant 0$' implies that $a(a - b)^2 + b(b - c)^2 + c(c - a)^2 \geqslant 0$. Unfortunately, the LHS of this inequality involves unwanted terms such as ab^2. However, you should be able to find a positive integer k such that $(ka + b)(a - b)^2$ does not involve ab^2. Then add three such terms to complete the solution.

(4) A third solution to (a) comes from pursuing the AM–GM idea more tenaciously. The LHS certainly invites you to use the AM–GM inequality: $(x + y + z)/3 \geqslant \sqrt[3]{xyz}$. Since each application of AM–GM will give a single term on the RHS, you must clearly use this idea ***ee times—first to obtain $a^2 b$ on the RHS, then to get the other two terms. Which three cubes x, y, and z should you multiply together if their cube root is to equal $a^2 b$ (that is, the product of the three cubes must be $a^6 b^3$)? Now rewrite the LHS in the form $(a^3 + a^3 + b^3)/3 + (b^3 + b^3 + c^3)/3 + (c^3 + c^3 + a^3)/3$ and complete the proof.

(5) One approach to part (b) is to multiply out the three brackets on the RHS and collect all terms on the LHS to show that the problem is equivalent to proving '$3abc + a^3 + b^3 + c^3 - (a^2 b + b^2 c + c^2 a) - (a^2 c + b^2 a + c^2 b) \geqslant 0$'. One can make this approach work, but it has to be handled carefully. (It is tempting to apply part (a) to show that the LHS is $\geqslant 3abc + a^3 + b^3 + c^3 - (a^2 c + b^2 a + c^2 b)$, and then to try to show that this simpler expression is $\geqslant 0$: unfortunately, this expression can be negative!)

(6) A much nicer approach is to assume (without loss of generality) that

$a \geqslant b$ and $a \geqslant c$, and to observe that then $a + b - c > 0$ and $c + a - b > 0$. If $b + c - a \leqslant 0$, then the RHS is also $\leqslant 0$, so the inequality holds. Thus you may assume that $b + c - a > 0$, and let $a + b - c = x$, $c + a - b = y$, and $b + c - a = z$. Then $(x + y)/2 = \underline{\quad}$, $(y + z)/2 = \underline{\quad}$, and $(z + x)/2 = \underline{\quad}$, and the inequality to be proved reduces to $((x + y)/2) \cdot ((y + z)/2) \cdot ((z + x)/2) \geqslant \sqrt{xy} \cdot \sqrt{yz} \cdot \sqrt{zx} = xyz$, which follows from three applications of the AM–GM inequality.

---

4. Given n points in space such that no plane passes through any four of them, let S be the set of all tetrahedra the vertices of which are four of the n points. Prove that a plane which does not pass through any of the n points cannot cut more than $n^2(n - 2)^2/64$ of the tetrahedra of S in quadrilateral cross-sections.

---

(1) It is often true that the worse a problem seems, the simpler it will be once you have made the effort to understand what is is saying. Imagine four points forming a tetrahedron in three dimensions. A plane which cuts off just one corner of the tetrahedron creates a **ia**u*a* cross-section. What can you say about a plane which cuts the tetrahedron in a *quadrilateral* cross-section? Where must the plane go?

(2) This (and one simple inequality) is all you need to solve the problem.

(a) Any plane which does not pass through any of the n points separates the n points into two sets—with i (say) on one side and $\underline{\quad}$ on the other side. Such a plane cuts a tetrahedron formed by four of the n points in a quadrilateral cross-section precisely when **o of the four points are on one side of the plane, and the other **o are on the other side. Since there are $\binom{i}{2}$ ways of selecting two points on one side and $\binom{n - i}{2}$ ways of selecting two points on the other side of the plane, the number of tetrahedra cut by this plane in quadrilateral cross-sections is exactly $\underline{\quad} \times \underline{\quad}$.

(b) Now $i(n - i) \leqslant ([i + (n - i)]/2)^2 = (n/2)^2$ (by the AM–GM inequality); similarly, $(i - 1)(n - i - 1) \leqslant \underline{\qquad\qquad}$. Hence $\binom{i}{2}\binom{n - i}{2} \leqslant \underline{\qquad\qquad}$.

---

5. Find, with proof, the smallest possible value of $|12^m - 5^n|$, where m and n are positive integers.

---

(1) This problem is unusual and may throw you at first; but it is an excellent challenge.

(a) When $m = n = 1$, $|12^m - 5^n| = $ ____ . Thus you know that the smallest possible value is at most ____ .

(b) Since 12 is e*e* and 5 is o**, the difference $|12^m - 5^n|$ is always o**. This reduces the possible smallest values to just *ou*; namely __, __, __, and 7.

(c) Extending the argument in (b), you know that 5^n is always a multiple of __, whereas 12^m is never a multiple of __. Hence the difference $|12^m - 5^n|$ can never be a multiple of __. Similarly, 12^m is always a multiple of 3, but 5^n is never a multiple of 3, so $|12^m - 5^n|$ is never a multiple of 3.

(d) Hence the smallest possible value of $|12^m - 5^n|$ is either __ or 7. You therefore have to decide whether you should try to find values m and n with $|12^m - 5^n| = 1$, or try to prove that none exist. Unless you can spot suitable values quickly, it makes sense to try to prove that none exist—for in the process of trying to prove that no values exist, you will find out more and more about what properties such values would have if they did exist.

(e) Suppose that $|12^m - 5^n| = 1$. Then $5^n \equiv \pm 1 \,(\mathrm{mod}\, 12)$, and $12^m \equiv \pm 1 \,(\mathrm{mod}\, 5)$. Show that $5^n \equiv -1 \,(\mathrm{mod}\, 12)$ has no solutions, and that $5^n \equiv 1 \,(\mathrm{mod}\, 12)$ precisely when n is e*e*. Show also that $12^m \equiv \pm 1 \,(\mathrm{mod}\, 5)$ if and only if m is e*e*. Thus $m = 2i$ and $n = 2j$ for some positive integers i and j. But then $\pm 1 = 12^m - 5^n = 12^{2i} - 5^{2j} = (12^i)^2 - (5^j)^2$ is a difference of two **ua*e*. However, the only squares which differ by 1 are __ and __. Use this to conclude that $|12^m - 5^n| \neq 1$.

---

6. Given distinct non-zero integers a_i $(1 \leqslant i \leqslant n)$, let p_i be the product of all the factors $(a_i - a_1), (a_i - a_2), \ldots, (a_i - a_n)$ except for the zero factor $(a_i - a_i)$. Prove that if k is a non-negative integer, $\sum_{i=1}^{n} a_i^k / p_i$ is an integer.

---

(1) This looks even worse than Question 2 above. But stay cool, calm and collected! To show that the given expression 'is an integer', you clearly have to put everything over a common denominator and explain why the denominator D must divide the numerator N. This does not look inviting, but it is not as bad as it may seem. However, it does require a willingness to treat the symbols a_i as 'algebraic indeterminates' or variables (rather than as integers with particular values), and to treat the expressions for D and N in the question as polynomials in these variables (rather than as integers), with 'division' meaning 'polynomial division'.

(2)(a) Suppose that you have to write the sum in the question as a rational expression with a common denominator. The simplest denominator to choose is just $D = p_1 p_2 \cdots p_n$. (You could try taking

$\sqrt{\left((-1)^{n(n-1)/2}p_1p_2\cdots p_n\right)}$, since every bracket $(a_i - a_j)$ occurs exactly twice, with opposite signs; however, square roots are algebraically messy.) The first two terms in the numerator N when the expression $\Sigma[(a_i^k)/p_i]$ is written with common denominator D are

$$(*) \qquad a_1^k(p_2p_3p_4\cdots p_n) + a_2^k(p_1p_3p_4\cdots p_n) + \ldots .$$

Write out the next two terms.

(b) Each factor in the product p_1 has the form $(a_1 - a_j)$ for some $j \neq 1$; in particular, each factor involves a_1, and one of these factors is $(a_1 - a_2)$. Similarly, each factor in the product p_2 involves a_2—and one of these factors is $(a_2 - a_1)$. No other product p_k involves the factor $(a_1 - a_2)$.

(c) Clearly, each term after the first two in $(*)$ is divisible by the product p_1p_2, and hence by $(a_1 - a_2)^2$. On the other hand, the sum of the first two terms in $(*)$ can be written as

$$p_3p_.\cdots p_n(a_1^k p_2 + a_2^k p_1)$$

and

$$a_1^k p_2 + a_2^k p_1$$
$$= (a_2 - a_1)\left[a_1^k(a_2 - a_3)\cdots(a_2 - a_n) - a_2^k(a_1 - a_3)\cdots(a_1 - a_n)\right].$$

The expression in square brackets on the RHS vanishes if $a_1 = a_2$; thus if you view this expression as a polynomial in a_1, then it must have $(a_1 - a_2)$ as a factor. Hence the numerator N is divisible by $(a_1 - a_2)^2$.

(d) Similarly, N is divisible by $(a_i - a_j)^2$ for each pair i, j. Hence, if you treat the symbols $a_1, a_2, \ldots, a_n$ as independent variables, and D as a polynomial in these n variables, then D divides N (as polynomials). If you then substitute the given integer values of $a_1, a_2, \ldots, a_n$ in the quotient N/D you obtain the (integer) quotient—hence proving that N/D is an integer.

16th British Mathematical Olympiad, 1980

1. Prove that the equation $x^n + y^n = z^n$, where n is an integer > 1, has no solution in integers x, y, and z with $0 < x \leqslant n$ and $0 < y \leqslant n$.

(1) This is a very special case of 'Fermat's Last Theorem'. The restrictions '$0 < x \leqslant n$, $0 < y \leqslant n$' must be important, but at first sight it is not at all clear

how they are meant to help. However, you should eventually stumble on the following idea.

(*)   Suppose (without loss of generality) that $x \leqslant y \leqslant n$, with $n \geqslant 2$.

Then $y < z$, so $z \geqslant y + 1$.

Hence

$$x^n + y^n \leqslant y^n + y^n \leqslant y^n + n \cdot y^{n-1} < (y+1)^n \leqslant z^n,$$

so there are no solutions.

---

2. Find a set S of seven consecutive positive integers for which a polynomial $P(x)$ of degree 5 exists with the following properties:
(a) all of the coefficients in $P(x)$ are integers;
(b) $P(n) = n$ for five numbers $n \in S$, including the least and the greatest;
(c) $P(n) = 0$ for some $n \in S$.

---

(1) You are given certain properties of the polynomial $P(x)$, but to determine a polynomial you would prefer to know its *oo**. Thus it makes sense to switch attention to the polynomial $Q(x) = P(x) - x$. This new polynomial $Q(x)$ has integer roots α, β, γ, δ, and ε, where $\alpha < \beta < \gamma < \delta < \varepsilon = \alpha + 6$, and satisfies $Q(a) = \underline{\quad}$, where a is one of the two integers a and a' different from β, γ, and δ which lie between α and ε. Hence

$$Q(x) = A(x - \alpha)(x - \beta)(x - \gamma)(x - \delta)(x - \varepsilon),$$

so $Q(a) = A(a - \alpha)(a - \beta)(a - \gamma)(a - \delta)(a - \varepsilon) = -a.$

Since a lies between α and $\varepsilon = \alpha + 6$, the possible values of the factors $a - \alpha$, $a - \beta$, $a - \gamma$, $a - \delta$, and $a - \varepsilon$ are highly restricted (being distinct integers between -5 and $+5$). For example, if $a = \alpha + 3$ and $a' = \alpha + 1$, the second condition becomes $Q(a) = A \cdot 3 \cdot 1 \cdot (-1)(-2)(-3) = -a$. Thus you may choose $A = 1$ and $a = 18$, whence $S = \{15, 16, 17, 18, 19, 20, 21\}$ and $P(x) = Q(x) + x$ is a monic polynomial. [Alternatively, you may prefer to choose $A = 18$ and $a = 1$, whence $S = \{-2, -1, 0, 1, 2, 3, 4\}$.]

---

3. Given a semi-circular region with diameter AB, P and Q are two points on the diameter AB, and R and S are two points on the semi-circular arc such that $PQRS$ is a square. C is a point on the semi-circular arc such that the areas of the triangle ABC and the square $PQRS$ are equal. Prove that a straight line passing through one of the points R and S and through one of the points A and B cuts a side of the square at the incentre of the triangle.

(1) Let the radius of the circle be 1; and let Q lie between P and B.

(a) Use Pythagoras to find the side length of the square $PQRS$ inscribed in the semicircle.

(b) Let CX be the perpendicular from C to AB. Equating areas shows that $(2/\sqrt{5})^2 = CX$. Thus C lies between B and R (or between A and S).

(c) BS passes through the incentre of triangle ABC if and only if BS is the *i*e**o* of $\angle ABC$. Thus it suffices to show that the chord AS has the same length as SC. Use Pythagoras to show that each of these chords has length $\sqrt{2 - (2/\sqrt{5})}$.

---

4. Find the set of real numbers a_0 for which the infinite sequence $\{a_n\}$ of real numbers defined by $a_{n+1} = 2^n - 3a_n$ $(n \geqslant 0)$ is strictly increasing.

---

(1) This is considerably tougher than the previous three questions. One approach is to find a closed formula for the nth term a_n. While it is clear that this will involve powers of 2 and 3, a beginner could be excused for not guessing that the general term a_n can be expressed in the form $a_n = A \cdot 2^n + B \cdot (-3)^n$.

(a) Use the recurrence relation to show that, if $a_n = A \cdot 2^n + B \cdot (-3)^n$, then $A = \underline{\quad}$.

(b) Use $a_1 = 2^0 - 3a_0$ to show that, if $a_n = A \cdot 2^n + B \cdot (-3)^n$, then $B = \underline{\quad}$.

(c) Prove (by induction) that $a_n = A \cdot 2^n + B \cdot (-3)^n$ for these values of A and B. Conclude that the sequence $\{a_n\}$ is increasing if and only if $a_0 = \underline{\quad}$.

(2) A more direct approach is to grind out successive values and to look for a closed form which depends directly on a_0. When doing this it is important *not* to multiply everything out, but to preserve the *form* of each term so that you can see what exactly is going on.

(a) $a_1 = 2^0 - 3^1 a_0$; $a_2 = 2^1 - 3(2^0 - 3a_0) = [2^1 \cdot 3^0 - 3^1 \cdot 2^0] + 3^2 a_0$. Write out similar expressions for a_3 and a_4.

(b) Write out the corresponding expression for a_{n+1}.

(c) The constant term '$2^n \cdot 3^0 - 2^{n-1} \cdot 3^1 + \ldots + (-1)^n 3^n \cdot 2^0$' is a GP with common ratio $r = \underline{\quad}$. Write down the formula for its sum. Simplify your expression as much as possible. Hence write down a closed form expression for the nth term

$$a_{n+1} = \underline{\qquad\qquad} + (-1)^{n+1} 3^{n+1} a_0.$$

(d) Suppose that $a_{n-1} < a_n < a_{n+1}$. Use your closed form expression for a_n to show that, if n is odd, the inequality $a_{n-1} < a_n$ implies that

$$a_0 < \frac{\dfrac{2^n + 3^n}{5} - \dfrac{2^{n-1} - 3^{n-1}}{5}}{3^{n-1} \cdot 4}$$

and that the inequality $a_n < a_{n+1}$ implies that

$$a_0 > \frac{\dfrac{2^n + 3^n}{5} - \dfrac{2^{n+1} - 3^{n+1}}{5}}{3^n \cdot 4}.$$

As n tends to infinity, the powers of 3 dominate, so the RHS of each inequality tends to ____. Thus there is only one possible value of a_0 which gives rise to an increasing sequence $\{a_n\}$; namely, $a_0 =$ ____. Check this starting value to show that it does indeed generate an increasing sequence.

---

5. In a party of ten people, you are told that among any three people there are at least two who do not know each other. Prove that the party contains a set of four people, none of whom knows the other three.

---

(1) Denote the ten people by the vertices of a network, or graph, joining two people precisely when they are acquainted. Given any vertex x, the first condition guarantees that none of the vertices joined to x can be joined to each other. If some vertex x were joined to four or more other vertices, you could choose any four of the vertices joined to x to obtain 'four persons none of whom knows the other three'.

(2) Thus you may assume that each person knows at most three people. The quickest way to finish the solution is then to observe that, given any vertex x, there are at least ____ vertices not joined to x. If we choose six vertices not joined to x, then by a well known result, among these six vertices there are either three vertices forming a triangle (that is, 'all of whom know each other'), or three vertices with no edges between them (that is, 'no two of whom know each other'). The first possibility is ruled out by the first condition in the question. Hence there must exist three vertices u, v, and w, not joined to x, no two of which are joined to each other—giving a set $\{u, v, w, x\}$ of 'four persons, none of whom knows the other three'.

15th British Mathematical Olympiad, 1979

---

1. Find all triangles ABC for which $AB + AC = 2$ and $AD + BC = \sqrt{5}$, where AD is the altitude through A, meeting BC at the point D.

(1) One of the difficulties in this problem is the fact that so many things are varying. So why not start by fixing the points B and C, and hence the length $BC = a$ (where $a < AB + AC = 2$). The second condition then becomes

$$AD = \sqrt{5} - a,$$

and so A lies on a line $*a*a**e*$ to BC and distance $\sqrt{5} - a$ away from it. Since B and C are fixed, the first condition now says that A lies on an $e**i**e$ with B and C as $*o*i$.

(a) Use the given information to write down the length _____ of the major axis of the ellipse.

(b) Show (using Pythagoras, or the eccentricity of the ellipse) that the minor axis has length $2\sqrt{1^2 - (a/2)^2}$. Conclude that $AD \leqslant \sqrt{1^2 - (a/2)^2}$.

(c) Finally, use the condition $AD = \sqrt{5} - a$ to show that $(a - (4\sqrt{5}/5))^2 \leqslant 0$.

Hence deduce that there is exactly one triangle satisfying the conditions in the question.

---

2. From a point O in three-dimensional space three given rays, OA, OB, and OC, emerge, with $\angle BOC = \alpha$, $\angle COA = \beta$, and $\angle AOB = \gamma$, $0 < \alpha, \beta, \gamma < \pi$. Prove that, given $2s > 0$, there are unique points X, Y, and Z on OA, OB, and OC respectively such that the triangles YOZ, ZOX, and XOY have the same perimeter $2s$. Express OX in terms of s, $\sin(\alpha/2)$, $\sin(\beta/2)$, and $\sin(\gamma/2)$.

It helps to draw a diagram that *looks* three-dimensional and that helps you see what is going on. What exactly must you do to prove that 'given $2s > 0$, there are unique points X, Y, and Z, and so on? Presumably you have to take *arbitrary* points X, Y, and Z; find expressions for the perimeters of triangles YOZ, ZOX, and XOY; and then equate these expressions and show that there is a unique

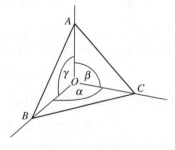

'solution'. Since XY can be expressed in terms of OX and OY (and γ), a good first move (in the spirit of *Name and Conquer*) is to let $OX = x$, $OY = y$, and $OZ = z$.

(a) Find an expression for XY^2 in terms of x, y, and γ.

(b) Suppose that triangle OXY has perimeter $2s$. Find another expression for XY^2.

(c) Equate the two expressions, and show that $(s - x)(s - y) = xy \sin^2(\gamma/2)$.

(d) Write down the corresponding expressions for $(s-y)(s-z)$ and for $(s-z)(s-x)$. Multiply the three expressions together and take the (positive) square root of both sides to obtain an equation for $(s-x)(s-y)(s-z)$. (When taking square roots you should explain why each of $(s-x)$, $(s-y)$, and $(s-z)$ is positive, and also why $\sin(\alpha/2)$, $\sin(\beta/2)$, and $\sin(\gamma/2)$ are all positive.) Finally, combine this equation with your equation for $(s-y)(s-z)$ to obtain an equation for $(s-x)$ in terms of x, $\sin(\alpha/2)$, $\sin(\beta/2)$, and $\sin(\gamma/2)$; hence show that 'given $2s > 0$, there is a unique' value of x (and similarly a unique value of y and z), so there are unique points X, Y, and Z as required.

---

3. $S = \{a_1, a_2, \ldots, a_n\}$ is a set of distinct positive odd integers. The differences $|a_i - a_j|$ $(1 \leqslant i < j \leqslant n)$ are all distinct. Prove that

$$\sum_{i=1}^{n} a_i \geqslant \tfrac{1}{3} n(n^2 + 2).$$

---

(1) Order the integers so that $a_1 < a_2 < \ldots < a_n$. You are told that the integers are all o**, so you can be sure that each of the differences $|a_i - a_j|$ is e*e*. Moreover, there are exactly ____ different pairs a_i, a_j $(1 \leqslant i < j \leqslant n)$. Since you are told that no two differences are equal, these $\binom{n}{2}$ even integers must all be different, so the largest difference $a_n - a_1$ must be greater than or equal to the $\binom{n}{2}$th even integer ____. Hence

$$a_n \geqslant a_1 + n(n-1) \geqslant 1 + n(n-1).$$

(2) You now know that $a_1 \geqslant 1$ and $a_n \geqslant n(n-1) + 1$. You need a lower bound for each a_i. Since the differences $|a_i - a_j|$ are all different,

$$a_2 \geqslant a_1 + (a_2 - a_1) \geqslant 1 + 2$$

(since $a_2 - a_1 \geqslant 2$). Similarly,

$$a_3 \geqslant a_1 + (a_2 - a_1) + (a_3 - a_2) \geqslant 1 + (2 + 4),$$

since $a_2 - a_1$ and $a_3 - a_2$ are positive, different, and even. In general,

$$a_r \geqslant a_1 + (a_2 - a_1) + \ldots + (a_r - a_{r-1}) \geqslant 1 + (2 + 4 + \ldots + 2(r-1)),$$
$$= 1 + r(r-1).$$

Hence

$$\sum_{i=1}^{n} a_i \geqslant \sum_{i=1}^{n} [1 + i(i-1)] \geqslant n + \sum_{i=1}^{n} i^2 - \sum_{i=1}^{n} i = \underline{\qquad}.$$

4. The function f is defined on the rational numbers and takes only rational values. For all rationals x and y,

(∗) $$f(x + f(y)) = f(x)f(y)$$

Prove that f is a constant function.

(1) Observe first that, if $f(x) = c$ is a constant function, then the LHS of the given functional equation (∗) is equal to ____, while the RHS is equal to ____^{2}. Only two rational numbers c satisfy $c = c^2$, namely, $c =$ ____ and $c =$ ____ . Thus you have to prove that the given functional equation implies either (a) $f(x) = 0$ for all rational x or (b) $f(x) = 1$ for all rational x.

(2) Suppose that $f(y) = 0$ for some rational y. Use (∗) to show that then $f(x) = 0$ for all rational x, so (a) holds.

(3) Thus you may assume from now on that $f(y) \neq 0$ for all y. In particular, $f(0) = c \neq 0$. It seems reasonable to try to prove first that $c = 1$, and then that '$f(x) = 1$ for all rational x'.

(i) Put $x = 0$ and $y = 0$ in (∗) to find the value of $f(c)$ in terms of c. Then put $x = c$ and $y = 0$ to find the value of $f(2c)$. Use a similar method to find $f(kc)$ ($k \in \mathbb{N}$).

(ii) You are told that 'f takes only rational values'. Hence $c = p/q$, for some $p \in \mathbb{Z}$, and some $q \in \mathbb{N}$, so $qc = p \in \mathbb{Z}$. Thus $f(p) = f(qc) = c^{q+1}$.

(iii) Put $x = -c$ and $y = 0$ in (∗) to find the value of $f(-c)$.

(iv) Put $x = 0$ and $y = -c$ in (∗) to find the value of $f(1)$. Then put $x = 1$ and $y = -c$ to find the value of $f(2)$. Let $n \in \mathbb{N}$ and suppose that you have already proved that $f(k) = c$ for all integers k, $-n < k \leqslant n$; put $x = \pm n$ and $y = -c$ to show that $f(-n) = c$ and $f(n + 1) = c$.

(v) Combine (ii) and (iv) to show that either $c =$ ____ , or q is e∗e∗ and $c =$ ____ . Then use (iii) to show that $c = -1$ is impossible.

(vi) Finally, let $r = m/n$, with $m \in \mathbb{Z}$ and $n \in \mathbb{N}$. Suppose that $f(r) = d = p'/q' \in \mathbb{Q}$. Imitate your method in (i) to show that $f(d) = d$, and that $f(kd) = d^k$ for all $k \in \mathbb{N}$. Conclude that $1 = f(p') = f(q' \cdot d) = d^{q'}$, so that $d = \pm 1$. Then imitate your method in (iii) to show that $f(-d) = 1/d$. Hence show that $d = 1$.

5. If n is a positive integer, denote by $p(n)$ the number of ways of expressing n as the sum of one or more positive integers. Thus $p(4) = 5$, as there are five different ways of expressing 4 in terms of positive integers; namely,

$$1 + 1 + 1 + 1, \quad 1 + 1 + 2, \quad 1 + 3, \quad 2 + 2, \quad \text{and} \quad 4.$$

Prove that $p(n + 1) - 2p(n) + p(n - 1) \geqslant 0$ for each $n > 1$.

(1) Write out the $p(3) = 3$ ways of expressing $n = 3$ as a sum. Do the same for $n = 4$ and $n = 5$. You have to find a reason which explains why '$p(5) - 2p(4) + p(3) \geqslant 0$' and which works equally well for other values of n.

(2) It is tempting to try to partition the set of expressions for $n + 1$ into three disjoint sets—two of which are easy to count and have sizes $p(n)$ and $p(n) - p(n - 1)$, in which case the third set need not be counted at all. The following two observations make this approach look even more promising.

(a) Explain why there are exactly $p(n)$ ways of writing $n + 1$ as a sum in which the smallest summand is a '1'.

(b) Use (a) to explain why the number of ways of writing n as a sum in which the smallest summand is '$\geqslant 2$' is $p(n) - p(n - 1)$.

Unfortunately, it is less clear how to fit these two tantalising facts together to complete the solution. (Each expression for n as a sum can be extended to an expression for $n + 1$ by adding a 1 at the front. Similarly, each exprssion for n as a sum with smallest summand $\geqslant 2$ can be extended to an exprssion for $n + 1$ by adding a 1 at the front. Unfortunately, the second set of expressions for $n + 1$ is a subset of the first, so all the expressions in (b) have been counted *twice*, while expressions which do not start with a '1' have not been counted at all.)

(3) If you look carefully at the failed idea in (2), you may see a way out. Let $p_1(n)$ denote the number of expressions for n with smallest summand '1', and let $p_2(n)$ denote the number of expressions for n with smallest summand $\geqslant 2$.

(a) Explain why $p_1(n) = p(n - 1)$. Hence explain why the number of expressions for $n + 1$ that have at least *two* summands equal to 1 is equal to $p(n - 1)$.

(b) Explain why $p_2(n) = p(n) - p(n - 1)$. Explain why this is equal to the number of expressions for $n + 1$ that have *exactly* one summand equal to 1.

(c) Finally, if you take each expression for n which has smallest summand $\geqslant 2$ and add one to its largest summand, you obtain exactly $p_2(n) = p(n) - p(n - 1)$ expressions for $n + 1$ in which the smallest summand is $\geqslant 2$.

(d) The construction in (c) gives some, but by no means all, of the expressions for $n + 1$ with smallest summand $\geqslant 2$: for example, when $n = 3$, it misses $2 + 2$. However, the three sets counted in (a), (b), and (c) are disjoint, so this gives the required inequality in the form

$$p(n + 1) \geqslant p(n - 1) + 2(p(n) - p(n - 1)).$$

(4) Alternatively, you may notice that the inequality to be proved can be written in the more natural form:

$$p(n + 1) - p(n) \geqslant p(n) - p(n - 1).$$

Thus you only need to show that, if $q(n) = p(n) - p(n - 1)$, then $q(n + 1) \geqslant q(n)$. By (2)(a), $q(n)$ is precisely the number of ways of writing n as a sum without using any 1's. To prove the inequality $q(n + 1) \geqslant q(n)$ you only have to show that there are at least as many ways of expressing $n + 1$ without using any 1's as there are ways of expressing n without using any 1's. This follows directly from (3)(c) (since each such sum for n can be changed into a sum for $n + 1$ by increasing the largest term by 1).

---

6. Prove that there are no prime numbers in the infinite sequence

$$10001, 100010001, 1000100010001, \ldots .$$

---

(1)(a) Any factorization of $10001 = a \times b$ with $a \geqslant b$ corresponds to writing 10001 as the difference of two squares $[(a + b)/2]^2 - [(a - b)/2]^2$. Moreover, since a and b are both odd and congruent (mod 4), $(a + b)/2$ is odd and $(a - b)/2$ is even. Checking $101^2, 103^2, 105^2, \ldots$ one soon finds that $10001 = 105^2 - \underline{\quad}^2$, giving the factorization $10001 = \underline{\quad} \times \underline{\quad}$. Hence the first term is not prime.

(b) The second term has digit sum equal to $\underline{\quad}$, is therefore divisible by $\underline{\quad}$, and so is also not a prime.

(c) The third term is visibly divisible by 10001. More generally, the $(mn - 1)$th term in the sequence is divisible by the $(m - 1)$th term, and so is not prime (if $m \geqslant 2$).

(2) This feels like progress. But it leaves the (still infinite!) subsequence of terms involving p 1's, where p is a prime $\geqslant 5$. So start again from scratch and think about the *form* of the terms under scrutiny.

(a) Let u_n denote the nth term (with $n + 1$ 1's). Write the fourth term as a sum of powers of 10: '$1 + \ldots$'. Then use the formula for the sum of a GP (with first term 1 and common ratio $r = \underline{\quad}$) to obtain a closed formula for the fourth term. Factorize the numerator as a difference of two squares, and so explain why u_4 is not prime.

(b) Use the same method to show that, in general, $u_n = [(10^{n+1})^4 - 1]/[10^4 - 1]$. Factorize the numerator as a difference of two squares. Hence show that, when $n \geqslant 2$, u_n is never prime.

14th British Mathematical Olympiad, 1978

---

1. Determine, with proof, the point P inside a given triangle ABC for which the product $PL \cdot PM \cdot PN$ is a maximum, where L, M, and N are the feet of the perpendiculars from P to BC, CA, and AB respectively.

---

(1) It is rather hard to see how the product $PL \cdot PM \cdot PN$ varies as P varies—there are just too many degrees of freedom! (This remains a problem even though the question implicitly suggests that the product will attain its maximum at some relatively familiar point P.) In such cases a good strategy is to treat one of the three variables, say PN, as fixed, and then:

(a) first maximize the value of the product for each fixed value of PN;

(b) then find the maximum of all these maxima as the value of PN varies.

(2) Draw a triangle ABC, mark a particular point P' inside $\triangle ABC$, and draw in the three perpendiculars $P'L'$, $P M'$, and $P'N'$. Let P be an *unknown* point in the triangle, and let PN be the perpendicular from P to AB.

(a) Draw the locus of points P for which $PN = P'N'$. Let this line cut CA in the point A', and cut CB in the point B'.

(b) Fix the value of PN and try to locate the point P on $A'B'$ for which $PL \cdot PM$ is a maximum. Use the fact that $\angle LB'P = \underline{\hspace{1cm}}$ to express PL in terms of PB' and $\angle B$; similarly, express PM in terms of PA' and $\angle A$. Use the fact that $\angle A$ and $\angle B$ are fixed (once $\triangle ABC$ is given) to conclude that $PL \cdot PM$ is a maximum precisely when $PB' \cdot PA'$ is a maximum. Finally, use the fact that, when PN is fixed, $PA' + PB' = \underline{\hspace{1cm}}$ is *o***a** to conclude that the maximum is attained at the *i**oi** of $A'B'$.

(c) Now let PN vary. For each value of PN the maximum value of the product $PL \cdot PM \cdot PN$ is attained at the midpoint of $A'B'$. These points all lie on the line joining C to the midpoint of AB. Hence the required point P lies on the *e*ia* from C to AB. Similarly, P must lie on the median from A to BC. Hence the maximum value of the product $PL \cdot PM \cdot PN$ is attained when P is at the *e***oi* of triangle ABC.

(3) Alternatively, $\Delta = \text{area}(ABC) = \frac{1}{2}PL \cdot a + \frac{1}{2}PM \cdot b + \frac{1}{2}PN \cdot c$. Hence $\Delta \geqslant \frac{3}{2}(abc)^{1/3}[PL \cdot PM \cdot PN]^{1/3}$ (by AM–GM), with equality if and only if $\text{area}(PBC) = \text{area}(PCA) = \text{area}(PAB) = \Delta/3$. Since Δ is fixed, it follows that $PL \cdot PM \cdot PN$ achieves its maximum value $(2\Delta/3)^3 \cdot (1/abc)$ when P is at the centroid.

2. Prove that there is no proper fraction m/n with denominator $n \leqslant 100$, the decimal expansion of which contains the block of consecutive digits '167', in that order.

(1) What a curious question! While there may be some clever way to solve it, the way it is stated leaves one with little

$$\begin{array}{r} ?\ ?\ ?\ ...\ \mathbf{1}\ \mathbf{6}\ \mathbf{7} \\ \hline n\,|\,m.\ 0\ 0\ 0\ ...\ {}^{a}0\ {}^{b}0\ {}^{c}0\ ... \end{array}$$

option but to try to extract some mathematically usable information from the long-division process of dividing n into m. Suppose that the usual process of dividing n into m were to give rise at some stage to the sequence '167'. You must show that n must then be greater than 100. Suppose that the remainders at the three crucial stages are a, b, and c as shown. Then $10a = n + \underline{\quad}$, $10b = 6n + \underline{\quad\quad}$, and $10c = 7n + d$ (where d is the remainder at the next stage, so $0 \leqslant d < n$).

(a) Substitute for b (in terms of n and a) from the first equation into the second. Then substitute for c from this new second equation into the third to show that $1000a = 167n + d$.

(b) Show that, if $n \leqslant 100$, then $a \leqslant \underline{\quad}$.

(c) Use $0 \leqslant d < n$ to show $\dfrac{167}{1000} \leqslant \dfrac{a}{n} < \dfrac{168}{1000}$.

(d) The most simple-minded (yet still fairly efficient) way to finish the job is to use the fact that $\frac{1}{6} < \frac{167}{1000}$ to exclude all the possible values of a, starting with $a = 16$. If $a = 16$, then

$$\frac{16}{96} = \frac{16}{(6 \times 16)} < \frac{167}{100} \leqslant \frac{16}{n} < \frac{168}{1000} < \frac{16}{95},$$

so that $95 < n < 96$, which is impossible. The other values of a may be excluded one at a time in the same way. However, the calculation for $a = 16$ effectively eliminates them all: since, if $a = i < 16$, then

$$\frac{i}{6i} < \frac{167}{1000} < \frac{a}{n} = \frac{i}{n} < \frac{168}{1000} < \frac{16}{95} < \frac{i}{6i - 1},$$

so that $6i - 1 < n < 6i$, which is impossible.

(2) An alternative to (1)(c), (d) is to notice that the equation $1000a = 167n + d$ implies that $-2a \equiv d \pmod{167}$. Thus $167 \mid d + 2a$, whereas $0 < d + 2a < 132$ (since $a \leqslant 16$, and d is a remainder on division by $n \leqslant 100$, so $d < 100$).
(3) You might like to refine the above methods to prove that the smallest

possible n the decimal of which contains the sequence '167' is $n = 131$ (with $m = 22$).

---

3. Show that there is one and only one sequence $\{u_n\}$ of integers such that

$$u_1 = 1, \quad u_1 < u_2, \quad \text{and} \quad u_n^3 + 1 = u_{n-1} u_{n+1} \quad \text{for all } n > 1.$$

---

(1) To determine that there is 'one and only one sequence $\{u_n\}$' you have to show two things: first, that any two such sequences must be identical; and, second, that there is at least one sequence with the given properties.

(a) Let $u_2 = a > 1$. Find expressions for u_3 and u_4. Explain why the sequence $\{u_n\}$ is completely determined once u_2 is given.

(b) You want $\{u_n\}$ to be a sequence of i**e**e**. Thus you may assume that a is an integer. Once $u_2 = a$ is chosen to be an integer, u_3 is automatically an integer. Look carefully at your expression for u_4. Show that for u_4 to be an integer a must divide ____. Hence conclude that $a =$ ____ (since $u_2 > u_1$). Thus any sequence of integers with $u_1 = 1$ and satisfying the given recurrence relation has to have $u_2 = a =$ ____, and so is uniquely determined.

(2) It remains to prove that the sequence obtained when $a = 2$ is in fact a sequence of integers. If you are tempted to calculate successive terms in the hope of spotting some pattern, you will soon have second thoughts! Thus the fact that u_{n-1} always divides $u_n^3 + 1$ must probably be proved algebraically. While there is endless scope here for going round and round in algebraic circles, the necessary calculation is not that hard.

(a) Direct calculation shows that $u_1 = 1$, $u_2 = 2$, $u_3 =$ ____, and $u_4 =$ ____.

(b) Suppose that you know that all terms up to (and including) u_n are integers, for some $n \geq 4$; in particular, $(u_{n-1}^3 + 1)/u_{n-2}$ is an integer. You have to prove that u_{n+1} is an integer; that is, you have to prove that the integer $u_n^3 + 1$ is divisible by the integer u_{n-1}.

$$u_n^3 + 1 = \frac{(u_{n-1}^3 + 1)^3}{u_{n-2}^3} + 1 = \frac{(u_{n-1}^3 + 1)^3}{(u_{n-1} u_{n-3} - 1)} + 1$$

$$= \frac{(u_{n-1}^3 + 1)^3 + (u_{n-1} u_{n-3} - 1)}{(u_{n-1} u_{n-3} - 1)}$$

$$\therefore \quad (u_n^3 + 1)(u_{n-1} u_{n-3} - 1) = (u_{n-1}^3 + 1)^3 + (u_{n-1} u_{n-3} - 1).$$

All terms in this final equation are integers (by induction). Moreover, the

RHS is an exact multiple of u_{n-1} (Why?), whereas the second factor on the LHS is relatively prime to u_{n-1} (Why?). Since the RHS (and hence the LHS) is divisible by u_{n-1}, it follows that u_{n-1} must divide the first factor $u_n^3 + 1$ on the LHS. Hence, $u_{n+1} = (u_n^3 + 1)/u_{n-1}$ is an integer.

---

4. An *altitude* of a tetrahedron is a straight line through a vertex which is perpendicular to the opposite face. Prove that the four altitudes of a tetrahedron are concurrent if and only if each edge of the tetrahedron is perpendicular to its opposite edge.

---

(1) You have two things to prove.

(a) First, when the altitudes are concurrent, you have to prove that each pair of opposite sides are 'perpendicular' to one another.

(b) Second, if each pair of opposite sides are 'perpendicular' to one another, you have to show that the altitudes are concurrent.

(2) Let the tetrahedron be $ABCD$. The repeated mention of 'altitudes' and 'perpendicular' tend to suggest trying a vector approach. Suppose first that the four altitudes (from A to BCD, from B to CDA, from C to DAB, and from D to ABC) all cross at the single point O. Take O as the origin, and let $\mathbf{a}$, $\mathbf{b}$, $\mathbf{c}$, and $\mathbf{d}$ be position vectors of A, B, C, and D.

(a) The altitude OA is perpendicular to the plane BCD and hence to any vector in that plane. Conclude that $\mathbf{a} \cdot (\mathbf{b} - \mathbf{c}) = \mathbf{a} \cdot (\mathbf{c} - \mathbf{d}) = \mathbf{a} \cdot (\mathbf{d} - \mathbf{b}) = 0$, and hence that $\mathbf{a} \cdot \mathbf{b} = \mathbf{a} \cdot \mathbf{c} = \mathbf{a} \cdot \mathbf{d}$.

(b) Repeat this argument for the other altitudes.

(c) Conclude that $(\mathbf{d} - \mathbf{a}) \cdot (\mathbf{c} - \mathbf{b}) = 0$, so that AD is perpendicular to BC. Show, similarly, that AB is perpendicular to CD, and that AC is perpendicular to BD.

(3) Now suppose that AD is perpendicular to BC, that AB is perpendicular to CD, and that AC is perpendicular to BD, and try to prove that the four altitudes are concurrent. Here you may choose between a *qualitative* argument ((a),(b)) and an explicit calculation ((c),(d)).

(a) Let A' be the foot of the perpendicular from A on to the opposite face BCD. Define B', C', and D' similarly. Since AB is perpendicular to CD, the perpendicular projection BA' of BA on to the plane BCD will also be perpendicular to CD. Thus the plane ABA' is perpendicular to CD, and so contains the altitude BB' (since BB' is the intersection of the plane ABA' and the plane through B perpendicular to AC). Hence the plane ABA' is the same as the plane BAB', and contains both AA' and BB'. Explain why these two lines cannot be parallel, and hence conclude that they cross—at some point O.

(b) Similarly, CC' meets AA'. Moreover, the point of intersection X of CC' and AA' lies on AA'—and hence in the plane ABA', and also on CC'—and hence in the plane CBC'. Thus X lies on the intersection of these two planes—namely on BB', as well as on AA'. Conclude that $X = O$. Hence complete the solution.

(c) If you prefer an explicit calculation, take the point A' as the origin of co-ordinates. Deduce that $\mathbf{a}\cdot\mathbf{b} = \mathbf{a}\cdot\mathbf{c} = \mathbf{a}\cdot\mathbf{d} = 0$. Then use the assumption $(\mathbf{d} - \mathbf{a})\cdot(\mathbf{b} - \mathbf{c}) = 0$ and so on to conclude that $\mathbf{d}\cdot\mathbf{b} = \mathbf{d}\cdot\mathbf{c} = \mathbf{b}\cdot\mathbf{c} = k$, say.

(d) Finally, show that there exists a value of t such that $t\mathbf{a}$ lies not only on the altitude AA' but also on the other three altitudes. For example, for $t\mathbf{a}$ to lie on BB' it must satisfy $(\mathbf{b} - t\mathbf{a})\cdot(\mathbf{d} - \mathbf{c}) = 0$, $(\mathbf{b} - t\mathbf{a})\cdot(\mathbf{c} - \mathbf{a}) = 0$, and $(\mathbf{b} - t\mathbf{a})\cdot(\mathbf{a} - \mathbf{d}) = 0$. Show that the first is automatically satisfied, and that the second holds precisely when $t = -(k/(\mathbf{a}\cdot\mathbf{a}))$. Then show that this value satisfies the third condition. Conclude that, since the value of t is independent of $\mathbf{c}$ and $\mathbf{d}$, the point $t\mathbf{a}$ will also lie on the other two altitudes.

---

5. Inside a cube of side 15 units there are 11 000 given points. Prove that there exists a sphere of unit radius containing at least six of the given points.

---

(1) It looks very much as though you will need to use the *pigeonhole principle*. If so, then you clearly have to cover the cube with sufficiently many spheres 'of unit radius', so as to guarantee that some sphere will contain *more than five* of the 11 000 points. If there were $\geqslant 2200$ unit spheres, then it is conceivable that each might contain no more than five of the points. Thus you need to cover the cube with fewer than 2200 unit spheres.

(2) You will find these large numbers impossible to handle unless you keep things extremely simple.

(a) What is the simplest way of cutting up a cube into lots of identical pieces with small 'diameter'?

(b) Suppose that you cut the large cube into lots of identical small cubes, each of side s, say. What is the radius of the smallest sphere containing one of these small cubes of side s? If this circumscribing sphere is to have radius 1, what is the largest possible value of s?

(3) If this approach is to work, you must:

(a) first cut up the 15 by 15 by 15 cube into identical smaller cubes with side length $\leqslant 2/\sqrt{3}$;

(b) then check that the number of small cubes is <2200—whence at least one of them will contain more than 5 of the 11 000 points.

(i) Show that $15/(2/\sqrt{3}) < 13$.

(ii) Check that $13^3 < 2200$.

(iii) Explain how this solves the problem.

---

6. Show that if n is a non-zero integer, $2\cos n\theta$ is a polynomial of degree n in $2\cos\theta$. Hence or otherwise, prove that if k is rational, then $\cos k\pi$ is either equal to one of the numbers 0, $\pm\frac{1}{2}$, or ± 1, or is irrational.

---

(1) The first part is a straightforward induction proof. When $n = 0$, $2\cos(0 \cdot \theta) = p_0(2\cos\theta)$, where $p_0(x) = 2$ is a polynomial of degree 0. When $n = 1$, $2\cos n\theta = p_1(2\cos\theta)$, where $p_1(x) = x$ is a *monic* polynomial of degree 1. Suppose that $2\cos n\theta = p_n(2\cos\theta)$, where $p_n(x)$ is a *monic* polynomial of degree n with integer coefficients. Use

$$\cos[(n+1)\theta] + \cos[(n-1)\theta] = 2\cos\underline{\qquad} \cdot \cos\underline{\qquad}$$

to show that $p_{n+1}(x)$ is the *monic* polynomial of degree $n+1$ (with integer coefficients) given by

$$p_{n+1}(x) = xp_n(x) - p_{n-1}(x).$$

(2) Now suppose that $k = p/q$ is rational ($p \in \mathbb{Z}, q \in \mathbb{N}, q \geqslant 2$). Let $x = 2\cos k\pi$. You have to prove that either $x \in \{-2, -1, 0, 1, 2\}$, or x is irrational.

(a) Show that $p_q(x) = 2\cos\underline{\qquad} = \pm 2$.

(b) Suppose that $x = r/s$ is rational ($r \in \mathbb{Z}, s \in \mathbb{N}, \mathrm{hcf}(r, s) = 1$). The leading term of $p_q(x)$ is then $x^q = (r/s)^q = r^q/s^q$, with $\mathrm{hcf}(r^q, s^q) = \underline{\qquad}$. Deduce that $p_q(x) = N/s^q$ for some integer N with $\mathrm{hcf}(N, s^q) = 1$. Finally, use (a) to conclude that $s = 1$. Hence complete the solution.

13th British Mathematical Olympiad, 1977

---

1. A non-negative integer $f(n)$ is assigned to each positive integer n in such a way that the following conditions are satisfied:

(a) $f(mn) = f(m) + f(n)$, for all positive integers m, n:

(b) $f(n) = 0$ whenever the units digit of n (in base 10) is a '3'; and

(c) $f(10) = 0$.

Prove that $f(n) = 0$, for all positive integers n.

(1) It is instructive (although not strictly necessary) to begin by considering all values of $n \leqslant 10$. You already know that $f(3) = 0$ and that $f(10) = 0$.

(a) Use '$f(mn) = f(m) + f(n)$, for all $m, n \geqslant 1$' to prove that $f(1) = 0$.

(b) Use $f(10) = 0$ to prove that $f(2) = 0$ and $f(5) = 0$.

(c) Use $f(2) = 0$ to prove that $f(4) = 0$.

(d) Use $f(2) = f(3) = 0$ to prove that $f(6) = 0$.

(e) Prove that $f(8) = 0$ and $f(9) = 0$.

(f) Now you know that $f(9) = 0$, use the fact that $f(63) = 0$ to prove that $f(7) = 0$.

(2) You are now ready to prove the '$f(n) = 0$' for all $n \geqslant 1$ by induction on n. The statement is certainly true when $n \leqslant 10$. Suppose that for some $n \geqslant 10$ you already know that $f(k) = 0$ for all $k \leqslant n$.

(a) If $n + 1$ can be factorized—say, $n + 1 = a \cdot b$, with $1 < a, b < n + 1$—use the fact that $f(a) = f(b) = 0$ to prove that $f(n + 1) = 0$.

(b) Suppose that $n + 1$ is prime. Then, since $n + 1 \geqslant 7$, the units digit of $n + 1$ is (i) 1, or (ii) 3, or (iii) ___, or (iv) ___. Hence some multiple $k(n + 1)$ has units digit equal to 3 (in case (i) take $k = 3$; in case (ii) take $k = 1$; in case (iii) take $k = 9$; in case (iv) take $k = 7$); hence $f(k(n + 1)) = $ ___ . Use this to prove that $f(n + 1) = 0$.

> 2. The sides BC, CA, and AB of a triangle touch a circle at X, Y, and Z respectively. Prove that the centre of the circle lies on the straight line through the midpoints of BC and of AX.

(1) The centre of the inscribed circle is precisely where the angle bisectors of $\angle A$ and $\angle B$ meet. Let the bisector of $\angle A$ meet BC at D. Let $\mathbf{a}$, $\mathbf{b}$, and $\mathbf{c}$ be the position vectors of the three vertices, and let a, b, and c be the lengths of the three sides (*not* the magnitudes of $\mathbf{a}$, $\mathbf{b}$, and $\mathbf{c}$). Let D have position vector $\mathbf{d}$.

(a) Show that $BD : DC = AB : AC$ (the 'Angle Bisector Theorem'). Use this to show that $\mathbf{d} = (b\mathbf{b} + c\mathbf{c})/(b + c)$.

(b) Now let the bisector of $\angle B$ in $\triangle ABD$ meet AD at I (the incentre of $\triangle ABC$). Let I have position vector $\mathbf{i}$. Use $BD = DC \cdot (AB/AC) = (a - BD) \cdot (c/b)$ to find BD. Then use the same method as in (a) to show that $\mathbf{i} = (a\mathbf{a} + b\mathbf{b} + c\mathbf{c})/(a + b + c)$.

(2) To simplify matters slightly, choose the origin at the midpoint L of BC. Then $\mathbf{c} = -\mathbf{b}$, so $\mathbf{i} = (a\mathbf{a} + (b - c)\mathbf{b})/(a + b + c)$.

(a) Show that $BX = (a + c - b)/2$. (Use $BX = BZ = u$, $CX = CY = v$, and $AY = AZ = w$, and solve $u + v = a$, $v + w = b$, and $w + u = c$ to find u.)

(b) Let X have position vector $\mathbf{x}$. Find the magnitude of $\mathbf{b}$ (in terms of a!). Hence show that $\mathbf{x} + ((a + c - b)/2)\cdot((2/a)\mathbf{b}) = \mathbf{b}$. Simplify to find $\mathbf{x}$ (in terms of $\mathbf{b}$ and a, b, and c).

(c) Let the midpoint T of AX have position vector $\mathbf{t}$. Show that $\mathbf{t} = \frac{1}{2}(\mathbf{a} + \mathbf{x}) = (\frac{1}{2})\mathbf{a} + \underline{\quad} \mathbf{b}$. Show that $(2a/(a + b + c))\mathbf{t} = \mathbf{i}$. Conclude that L, I, and T are collinear.

(3) If 'the side BC' refers to the doubly infinite line BC (rather than the line *segment BC*), then the circle may be an *escribed* circle. You should check that a similar calculation works in this case.

---

3. (a) Prove that if x, y, and z are non-negative real numbers, then
$$x(x - y)(x - z) + y(y - z)(y - x) + z(z - x)(z - y) \geqslant 0.$$

(b) Hence or otherwise, show that, for all real numbers a, b, and c,
$$a^6 + b^6 + c^6 + 3a^2b^2c^2 \geqslant 2(b^3c^3 + c^3a^3 + a^3b^3).$$

---

(1) The simplest, but least elegant, solution to (a) is probably to use the symmetry of the LHS to observe that one may assume $z \geqslant y \geqslant x \ (\geqslant 0)$.

(a) Show that $x(x - y)(x - z) \geqslant 0$.

(b) Check that $z(z - x)(z - y) \geqslant 0 \geqslant y(y - z)(y - x)$.

(c) Use $z \geqslant y \geqslant 0$ (and hence $z - x \geqslant y - x$ to show that
$$z(z - x)(z - y) + y(y - z)(y - x) = (z - y)[z(z - x) - y(y - x)] \geqslant 0.$$

(d) Combine (a) and (c) to prove the required inequality.

(Note that (6) in the outline solution to 1981 Question 3(b) shows that the two questions are exactly equivalent.)

(2) Since the terms on the LHS of (a) are of degree 3, it is natural in part (b) to start by putting $x = a^2$, $y = \underline{\quad}$, and $z = \underline{\quad}$ in the inequality of part (a) to obtain an inequality involving $a^6 + b^6 + c^6 + 3a^2b^2c^2$.

(a) Write out the resulting inequality.

(b) Thus, to prove the inequality in part (b) it suffices to prove
$$a^4(b^2 + c^2) + b^4(c^2 + a^2) + c^4(a^2 + b^2) \geqslant 2(b^3c^3 + c^3a^3 + a^3b^3).$$
Write $a^4(b^2 + c^2) + b^4(c^2 + a^2) + c^4(a^2 + b^2) - 2(b^3c^3 + c^3a^3 + a^3b^3)$ as the sum of three perfect squares, and so complete the solution.

> 4. The equation $x^3 + qx + r = 0$, where $r \neq 0$, has roots u, v, and w. Express the roots of the equation $r^2x^3 + q^3x + q^3 = 0$ in terms of u, v, and w. Show that if u, v, and w are real, then this latter equation has no root in the interval $-1 < x < 3$.

(1) This question suggests that, though the two equations look different, they are in fact closely related. If the equations are related in a simple way (say with x in the original equation corresponding to kx in the second), then they should at least have the same *o***a** term.

(a) Multiply the second equation by a constant factor to give them both the same constant term r.

(b) Then identify the change of variable $y = kx$ which turns the second equation into $y^3 + qy + r = 0$. Hence express the roots of the second equation in terms of u, v, w, q, and r.

(2) The question asks you to 'express the roots of the second equation in terms of u, v, and w' only. Therefore you have to find an expression for q/r in terms of u, v, and w. So far you have made no use of the standard relations between the roots u, v, and w and the coefficients 1, 0, q, and r.

(a) Use the fact that the coefficient of x^2 is zero to write down a relation which u, v, and w must satisfy.

(b) Use the known values of the other coefficients in the original equation to write down relations of the form $uv + vw + wu =$ ____ and $uvw =$ ____. Divide the first of these two relations by the second and simplify to obtain an expression for q/r in terms of u, v, and w. Hence express the three roots of $r^2x^3 + q^3x + q^3 = 0$ in terms of u, v, and w only.

(3) Now suppose that u, v, and w are all real. You have to find some (preferably simple) way to use this additional information to prove that none of the roots of the second equation lie in the interval $-1 < x < 3$.

(a) If u, v, and w are all real, explain why $f(x) = x^3 + qx + r$ must have a maximum at $x = -\sqrt{(-q/3)}$ and a minimum at $x = \sqrt{(-q/3)}$. Conclude that $q < 0$.

(b) Explain why the maximum must be above the y-axis and the minimum must be below the y-axis. Try to simplify the two conditions '$f(-\sqrt{(-q/3)}) > 0$' and '$f(\sqrt{(-q/3)}) < 0$'. This leads to rather messy algebra. You may have to pursue this; but it is worth looking for a simpler alternative. In particular, you should be aware of the fact that you have still not made use of the important relation (2)(a)!

(c) What does the given information '$r \neq 0$' tell you about the roots u, v, and w?

(d) If u, v, and w are all real, how many of them can be negative? (Use the relation that you found in (2)(a).) Conclude that you may assume that u is negative, that w is positive, and that $u < v < w$, with $v \neq 0$.

(e) Suppose that $u = -(v + w)$ is the only negative root. Then $r > 0$ (since $uvw =$ ____). Hence qu/r is $*o*i*i*e$, whereas the other two roots qv/r and qw/r of the second equation are $*e*a*i*e$, with $qw/r < qv/r$. Moreover,

$$(q/r)\cdot u = -(1/u + 1/v + 1/w)\cdot u = -1 + (v + w)/v + (v + w)/w$$

$$= 1 + w/v + v/w \geqslant \underline{\quad}$$

(since $((v/w) - 1)^2 \geqslant 0$ can be rearranged to show that $w/v + v/w \geqslant 2$). Similarly,

$$qw/r < qv/r = -(1/u + 1/v + 1/w)\cdot v$$

$$= -1 + (1/(v + w) - 1/w)\cdot v < -1 + 0$$

(since $v + w > v > 0$ implies that $(1/(v + w) - 1/v) < 0$).

(f) The case in which u and v are both negative requires only a slight modification of the above argument. (However, this is not strictly necessary, since if u, v, and w are roots of $x^3 + qx + r = 0$ with u and v negative, then $-u$, $-v$, and $-w$ are roots of $x^3 + qx - r = 0$, and both equations give rise to exactly the same 'second' equation $r^2 x^3 + q^3 x + q^3 = 0$.)

---

5. The regular pentagon $A_1 A_2 A_3 A_4 A_5$ has sides of length $2a$. For each $i = 1, 2, \ldots, 5$, K_i is the sphere with centre A_i and radius a. The spheres $K_1, K_2, \ldots, K_5$ are all touched externally by each of the spheres P_1 and P_2—also of radius a. Determine, with proof, whether or not P_1 and P_2 have a common point.

---

(1) This question is slightly ambiguous. Imagine that the regular pentagon $A_1 A_2 A_3 A_4 A_5$ lies in a horizontal plane Π. If the spheres K_1, K_2, K_3, K_4, and K_5 are touched externally by each of two spheres P_1 and P_2, then one of these spheres—P_1, say—must touch the spheres K_i above the plane Π and the other—P_2, say—must touch the spheres K_i below the plane Π.

If the spheres P_1 and P_2 were *solid*, then they could not interpenetrate each other, so the only way in which they could 'have a common point' would be if the lowest point X of P_1 (and hence the highest point of P_2) were to lie exactly at the circumcentre C of the pentagon $A_1 A_2 A_3 A_4 A_5$. Let O_i be the centre of P_i. If $X = C$, then $O_1 C = O_1 X =$ ____ (given); since $A_1 O_1 =$ ____ (as the two spheres K_1 and P_1 touch), and $\angle A_1 C O_1 =$ ____, this would imply that the circumradius $r = A_1C$ of the pentagon $A_1A_2A_3A_4A_5$ is equal to $(\sqrt{3})a$, which is easily seen to be false (for example, since $\sin 36° \neq 1/\sqrt{3}$).

(2) One suspects that candidates were supposed to consider the spheres P_1 and P_2 as *loci* which may interpenetrate. The question then becomes whether the lowest point X of the sphere P_1 lies above or below (or in) the plane Π: that is, whether $O_1X > O_1C$ or $O_1X < O_1C$ (or $O_1X = O_1C$). One way to decide is to use the same right-angled triangle A_1CO_1 to calculate $CO_1 = \sqrt{(4a^2 - \rho^2)}$. You must then decide whether (i) $\sqrt{(4a^2 - \rho^2)} < a$ or (ii) $\sqrt{(4a^2 - \rho^2)} > a$ (or $\sqrt{(4a^2 - \rho^2)} = a$).

(a) Let M be the midpoint of A_1A_2. Use $\triangle A_1MC$ to show that $\rho = a/\sin 36°$.

(b) Show that (i) holds if $\sin 36° < 1/\sqrt{3}$, and that (ii) holds if $\sin 36° > 1/\sqrt{3}$.

(c) Use the geometry of the regular pentagon to show that $\sin 36° = (1 + \sqrt{5})/4$. Hence solve the problem.

---

6. The polynomial $26(x + x^2 + x^3 + \dots + x^n)$, where $n > 1$, is to be decomposed into a sum of polynomials, not necessarily all different. Each of these polynomials is to be of the form $a_1x + a_2x^2 + a_3x^3 + \dots + a_nx^n$, where each a_i is one of the numbers $1, 2, 3, \dots, n$ and no two a_i are equal. Find all the values of n for which this decomposition is possible.

---

(1) The wording here is very strange; but once you manage to understand what it is saying, the problem is actually rather easy. You have to find all n for which the given polynomial

$$26(x + x^2 + x^3 + \dots + x^n)$$

can be written as a sum of other polynomials of the form

$(*)$ $\qquad a_1x + a_2x^2 + a_3x^3 + \dots + a_nx^n,$

in which the coefficients a_i are just the integers $1, 2, 3, \dots, n$ in some order.

(a) You are not told exactly what these other polynomials look like. However, you know that the largest coefficient in $(*)$ is always ____, and that each of the coefficients in the original polynomial = ____. Deduce that $n \leqslant$ ____.

(b) Explain why $n = 26$ is impossible. Show that $n = 25$ is possible.

(2) This certainly feels like progress, but you need another idea if you are to avoid the mess of trying to deal with each of the remaining possible values of n in turn.

(a) What is the sum of all the coefficients in (∗)? If you were to add up k polynomials of this form, what would be the sum of all the coefficients in the resulting polynomial?

(b) Suppose that it is possible to find k polynomials like the one in (∗), the sum of which is equal to the polynomial $26(x + x^2 + x^3 + \ldots + x^n)$. Show that $kn(n + 1)/2 = 26n$. Use the fact that $n > 1$ and $n <$ ___ (from (1)) to conclude that $n =$ ___ (with $k =$ ___), or $n =$ ___ (with $k =$ ___), or $n = 25$ (with $k = 2$).

(c) You already know that $n = 25$ works. Now take the other two possibilities in turn and show that they also work.

12th British Mathematical Olympiad, 1976

> 1. Find, with proof, the length d of the shortest straight line which bisects the area of an arbitrary given triangle, but which does not pass through any of the vertices. Express d in terms of the area Δ of the triangle and one of its angles. Show that there is always a shorter line (not straight) which bisects the area of the given triangle.

(1) Let the triangle be ABC. The *median* joining A to the midpoint L of BC splits the triangle into two parts of equal area (equal bases, same height). The same is true for the other two medians. To confront a common misconception, it is instructive to begin by proving that these are the *only* lines through the centroid G that bisect the area of the triangle!

(a) Suppose that y ($\neq A$) is a point on AB and that Z is a point on BC such that Y, G, and Z are collinear. Let u, v, w, and x denote the areas of AYG, $ABZC$, ZGL, and $BLGY$. Show that YZ bisects the area of the triangle if and only if $u = w$. Let CN be the median from C to AB; use $GC = 2 \cdot GN$ to deduce that $GZ = 2 \cdot GY$.

(b) Hence prove that $Z = C$. (Suppose that $Z \neq C$. Show that $\triangle GNY$ and $\triangle GCZ$ would then be similar. Conclude that $\angle GYN = \angle GZC$, whence YN would have to be parallel to CZ.)

(2) Choose points P on AB and Q on AC such that PQ bisects the area of $\triangle ABC$. Let AP and AQ have lengths p and q respectively. Let $PQ = x$.

(a) Use the cosine rule in $\triangle APQ$ to express x^2 in terms of p, q, and $\angle A$. You want to find the minimum possible value of x^2.

(b) Express the area of $\triangle APQ$ in terms of p, q, and $\angle A$. Put this equal to $\Delta/2$, and hence show that, as P and Q vary (with PQ still bisecting the area of $\triangle ABC$), the product pq remains constant.

(c) Combine (b) and (a) to deduce that x^2 is a minimum precisely when $p^2 + q^2$ is a minimum. Next explain why $p^2 + q^2$ is a minimum precisely when $(p - q)^2$ is minimum. Hence prove that, if $\sqrt{(\Delta/\sin A)}$ is less than or equal to both b and c, then the minimum occurs when $p = q = \sqrt{(\Delta/\sin A)}$.

(d) Show that when $p = q$, $x^2 = 2p^2(1 - \cos A) = 4p^2 \sin^2(A/2)$. Conclude that the required length d is equal to the minimum of $\sqrt{2\Delta \tan(A/2)}$, $\sqrt{2\Delta \tan(B/2)}$, and $\sqrt{2\Delta \tan(C/2)}$ —and so occurs for the smallest of the three angles A, B, and C. You should check that if $a \leqslant b$ and $a \leqslant c$, then $\sqrt{(\Delta/\sin A)} < b$ and $\sqrt{(\Delta/\sin A)} < c$. (The strict inequality here implies that the shortest straight area-bisector is never a median.) This completes the first part of the question.

(3) To find a 'shorter (not straight) line' which bisects the area of $\triangle ABC$, suppose that $\angle BAC = \theta$ is the smallest angle of $\triangle ABC$.

(a) Choose X on BC such that AX bisects $\angle BAC$ and apply the same reasoning as in (2) to $\triangle ABX$ and $\triangle AXC$ separately. The shortest line bisecting the area of $\triangle ABX$ has length d_1 and the shortest line bisecting the area of $\triangle AXC$ has length d_2. Prove that $d_1 + d_2 < d$. Use the result of (2)(c) to explain why the cut of length d_1 has an endpoint in common with the cut of length d_2.

(b) If you find proving $d_1 + d_2 < d$ slightly messy, you may prefer the following argument. If you bisect $\angle BAX$ and $\angle XAC$ and repeat the above procedure, you obtain a (not straight) line which consists of four straight segments, and which again bisects the area of the triangle. Repeating this over and over again gives a sequence of polygonal lines, which bisect the area of $\triangle ABC$, and which approximate more and more closely to an arc of a *i***e with centre A, radius $r = \sqrt{(\Delta/\theta)}$ (where θ is measured in radians), and length $r \cdot \theta = \sqrt{\Delta \theta}$. Show that $\sqrt{\Delta \theta} < \sqrt{2\Delta \tan(\theta/2)}$. Hence complete the second part of the question.

---

2. Prove that if x, y, and z are positive real numbers, then
$$\frac{x}{y+z} + \frac{y}{z+x} + \frac{z}{x+y} \geqslant \frac{3}{2}.$$

---

(1) There are many approaches here that work (and even more that simply go round in circles). Two things are worth noticing. First, you would prefer not to have those mixed sums '$y + z$' and so on in the denominator. Second, the LHS of the inequality to be proved is unchanged when you replace x, y,

and z by λx, λy, and λz for any $\lambda \geqslant 0$; you may therefore choose a scaling factor to suit yourself.

(a) Multiply all three variables by '$1/(x+y+z)$', to obtain new positive variables—which will also be called x, y, and $z(!)$—with sum $x+y+z=$ ____.

(b) Rewrite the LHS in the form '$[x/(1-x)]+\ldots$'. Then write $x/(1-x)=-1+$ ____ and show that the inequality to be proved is equivalent to

$$\frac{\dfrac{1}{1-x}+\dfrac{1}{1-y}+\dfrac{1}{1-z}}{3} \geqslant \frac{3}{2}.$$

(c) The LHS now invites you to use the AM–GM inequality. It would therefore suffice to prove that

$$\sqrt[3]{\frac{1}{(1-x)(1-y)(1-z)}} \geqslant \frac{3}{2}.$$

(d) Notice that $(1-x)+(1-y)+(1-z)=$ ___, so (by the AM–GM inequality) $\sqrt[3]{(1-x)(1-y)(1-z)} \leqslant$ ___, which completes the solution.

    [A different way of getting rid of the mixed sums in the denominators is to put $u=y+z$, $v=z+x$, and $w=x+y$. Then $x=(v+w-u)/2$ and so on, so

$$\frac{x}{y+z}+\frac{y}{z+x}+\frac{z}{x+y}=\frac{1}{2}\left(\frac{v}{u}+\frac{u}{v}+\frac{w}{v}+\frac{v}{w}+\frac{u}{w}+\frac{w}{u}-3\right)$$

$$\geqslant \tfrac{1}{2}(2+2+2-3).]$$

(2) Alternatively, let $X=(x+y+z)/(y+z)$, $Y=(x+y+z)/(z+x)$, and $Z=(x+y+z)/(x+y)$. Then the *harmonic mean* H of X, Y, and Z satisfies

$$H=\frac{3}{\dfrac{1}{X}+\dfrac{1}{Y}+\dfrac{1}{Z}}=\underline{\quad}.$$

The standard harmonic mean inequality says that H is less than or equal to the arithmetic mean; hence $\tfrac{3}{2} \leqslant (X+Y+Z)/3$, so $X+Y+Z \geqslant \tfrac{9}{2}$, which is precisely the inequality to be proved.

(3) An even simpler method is to multiply out and show (using $x, y, z > 0$) that the inequality to be proved is equivalent to

$$2(x^3+y^3+z^3) \geqslant x^2y+y^2z+z^2x+xy^2+yz^2+zx^2.$$

Now use 1981 Question 3(a) (twice).

> 3. $S_1, S_2, \ldots, S_{50}$ are subsets of a finite set E. Each subset contains more than half the elements of E. Show that it is possible to find a subset F of E having not more than five elements, such that each S_i $(1 \leqslant i \leqslant 50)$ has an element in common with F.

(1) Suppose that the set E has n elements. Count the number of pairs (x, i) in the set $\mathscr{P} = \{(x, i): x \in S_i\}$ in two different ways.

(a) Suppose that, on average, each element x belongs to A of the sets S_i. There are exactly __ possible first co-ordinates x, and each element x occurs, on average, in __ pairs (x, i). Hence there are $n \cdot A$ such pairs.

(b) On the other hand, there are __ possible second co-ordinates i, and each set S_i has more than $n/2$ elements; hence there are $> 50 \cdot (n/2)$ pairs.

(c) Conclude that $A > 25$, and hence that some element x_1 must belong to at least 26 of the sets S_i.

(2) Removing 26 of the sets S_i which contain the element x_1 leaves exactly __ sets S_j. As before, count pairs (x, j) in two ways, where $x \in S_j$ and S_j is one of the remaining 24 sets. Conclude that some element x_2 must belong to at least 13 of the remaining sets S_j.

(3) Removing 13 of the remaining sets that contain the element x_2 leaves exactly __ sets. The same argument shows that some element x_3 must belong to at least six of the remaining sets.

(4) Removing six of the remaining sets that contain x_3 leaves just __ sets. The same argument then shows that some element x_4 must belong to at least three of the remaining sets. Finally, each of the two other sets contains more than half the elements in E, so they have at least one element x_5 in common. The set $F = \{x_1, x_2, x_3, x_4, x_5\}$ then does the trick.

> 4. Prove that if n is a non-negative integer, then $19 \times 8^n + 17$ is not a prime number.

(1) Let $19 \cdot 8^n + 17 = u_n$.

(a) If $n = 0$, then $u_0 = 19 \cdot 8^0 + 17 =$ _____ . Factorize this!

(b) If $n = 1$, then $u_1 = 19 \cdot 8^1 + 17 =$ _____ . Factorize this!

(c) If $n = 2$, then $u_2 = 19 \cdot 8^2 + 17 =$ _____ . Factorize this!

(2) There is little point doing further computations until you have an idea to test (and if the idea is a mathematical one, there may well be better ways to test it than using brute force computation). The values of u_0 and u_1 are misleading in that they are both **ua*e*, whereas u_2 is not. However, it is

hard not to notice that the only small prime factor of u_2 is ___, and that this is also a factor of u_0.

(a) Show that, if n is e∗e∗, then $8^n \equiv$ ___ (mod 3). Hence show that $19 \cdot 8^n + 17$ is never prime when n is even.

(b) In the same (optimistic) spirit, the only small prime factor of u_1 is ___. Show that, if n is o∗∗, then $8^n \equiv \pm$ ___ (mod 13). Hence show that $19 \cdot 8^n + 17$ is never prime when n is an odd number of the form ___.

(c) You are clearly going to have to work a little harder when $n = 4k + 3$. However, if you consider which small primes p might divide $u_3 = 19 \cdot 8^n + 17$, then it should not take you long to exclude $p = 2$ and $p = 3$, and hence to discover that u_3 is a multiple of ___. You should then be able to prove that $19 \cdot 8^n + 17$ is a multiple of ___ whenever n has the form $4k + 3$.

---

5. Prove that, if α and β are real numbers, and r and n are positive integers with r odd and $r \leqslant n$, then

$$\sum_{t=0}^{(r-1)/2} \binom{n}{r-t}\binom{n}{t}(\alpha\beta)^t(\alpha^{r-2t} + \beta^{r-2t})$$

$$= \sum_{t=0}^{(r-1)/2} \binom{n}{r-t}\binom{r-t}{t}(\alpha\beta)^t(\alpha + \beta)^{r-2t}.$$

---

(1) The easiest approach (that is, one which allows most students to get started) almost never leads to the easiest, or most satisfying, solution. In this problem, perhaps the most straightforward approach (although one which will test your ability to manipulate algebraic expressions correctly) is simply to find the coefficient of α^s (say) on each side, and then to show that corresponding coefficients are equal. (A neater method is to try to identify *what it is that both sides are equal to*: see (3).)

(2)(a) For $s \leqslant (r-1)/2$, the coefficient of α^s on the LHS is $\binom{n}{r-s}\binom{n}{s}\beta^{r-s}$, and the coefficient of α^{r-s} is _____.

(b) The coefficient of α^s on the RHS is more complicated, in that there is a contribution from each 'term' in the sum. Let $s \leqslant (r-1)/2$. Then $\alpha^s = \alpha^t \cdot \alpha^{s-t}$ for each t, $0 \leqslant t \leqslant s$. Check that the coefficient of α^s on the RHS is

$$\sum_{t=0}^{s} \binom{n}{r-t}\binom{r-t}{t}\beta^t\binom{r-2t}{s-t}\beta^{r-t-s}.$$

(c) Show   that

$$\binom{n}{r-t}\binom{r-t}{t}\binom{r-2t}{s-t} = \binom{n}{t}\binom{n-t}{s-t}\binom{n-s}{r-t-s} = \binom{n}{s}\binom{s}{t}\binom{n-s}{r-t-s}.$$

(d) Show that $\displaystyle\sum_{t=0}^{s}\binom{s}{t}\binom{n-s}{r-t-s} = \binom{n}{r-s}$. (Suppose that you are given a
set N of size n and a fixed subset $S \subseteq N$ of size s. What does the RHS
of this identity count? Now explain why the LHS counts the same
collection of sets in a different way.) Hence solve the problem.

(3) An alternative approach comes from observing that $\binom{n}{t}$ is the coefficient

of α^t in $(1+\alpha)^n$, and that $\binom{n}{r-t}$ is the coefficient of β^{r-t} in $(1+\beta)^n$.
Hence the tth summand on the LHS of the identity in the question consists
of precisely those terms in the expansion of $(1+\alpha)^n(1+\beta)^n$ which have
degree r. Another way of calculating the terms of degree r in the expansion
is to multiply $(1+\alpha)(1+\beta)$ first and then to expand the product $(1+
(\alpha+\beta)+\alpha\beta)^n$. In multiplying out the n brackets $(1+(\alpha+\beta)+\alpha\beta)$, for
each term one may choose either a '1', or an '$\alpha+\beta$', or an '$\alpha\beta$' from each
bracket. Choosing '$\alpha\beta$' from t brackets contributes a factor $(\alpha\beta)^t$ of degree
$2t$; thus, to land up with a term of degree r, it remains to choose '$\alpha+\beta$' from
exactly $r-2t$ brackets (and '1' from the remaining $n-r+t$ brackets). There
are $\binom{n}{r-t} = \binom{n}{n-r+t}$ ways of choosing the $n-r+t$ brackets from which
to take a '1'; there are then $\binom{r-t}{t}$ ways of choosing t of the remaining $r-t$
brackets from which to take '$\alpha\beta$' (with '$\alpha+\beta$' taken from the other $r-2t$
brackets). Hence the tth summand on the RHS also gives the terms of degree
r in the expansion of $(1+(\alpha+\beta)+\alpha\beta)^n$, so the two sides are equal.

---

6. A sphere with centre O and radius r is cut by a horizontal plane
   distance $r/2$ above O in a circle K. The part of the sphere above the
   plane is removed and replaced by a right circular cone having K as its
   base and having its vertex V at a distance $2r$ vertically above O. Q is a
   point on the sphere on the same horizontal level as O. The plane OVQ
   cuts the circle K in two points X and Y, of which Y is the further from
   Q. P is a point of the cone lying on VY, the position of which can be
   determined by the fact that the shortest path from P to Q over the
   surfaces of cone and sphere cuts the circle K at an angle of $45°$. Prove
   that $VP = \sqrt{3}\cdot r/\sqrt{(1+1/\sqrt{5})}$.

---

(1) Draw a decent diagram (more or less to scale). Your diagram should
suggest something slightly unexpected. Show that VY is tangent to the sphere
at Y.

(2) Position V directly above the 'north pole'. Then K becomes a circle of latitude. Let PQ cut the circle K at T.

(a) Explain why QXT is *not* a 'spherical triangle'.

(b) Let the circle of longitude through T meet the equator at Z. Check that QZT is a spherical triangle. Let q, z, and t denote the angles subtended by the arcs ZT, TQ, and QZ at the centre of the sphere. Use the given spherical triangle formulae to show that $\sqrt{2} \cdot \sin t = \sin z$ and that $(\sqrt{3}/2)\cos t = \cos z$. Conclude that $\sin t = 1/\sqrt{5}$.

(c) Check that one half of the great circle through Q and T starts at Q, curves up to T, and rises further above the circle of latitude K before cutting K again at a point T', and then dropping to Q' opposite Q on the sphere. Conclude that the angle QOZ is acute. Hence show that the arc YT on the circle K has length $(\sqrt{3}/2)r(\pi - \arcsin(1/\sqrt{5}))$. (This tells you exactly where the point T is on the base K of the cone.)

(3) Slit the cone along VY and lay it flat.

(a) Show that $VY = \underline{\quad}$.

(b) Use the known length of the arc YT on the circle K to calculate $\angle YVT$. Then use the fact that $\angle VTP = \pi/4$ to find $\angle VPT$.

(c) Let $2\theta = (\arcsin(1/\sqrt{5}))$. Use the double angle formulae to show that $\cos^2\theta = (5 + 2\sqrt{5})/10$ and $\sin^2\theta = (5 - 2\sqrt{5})/10$.

(d) Use the (ordinary) sine rule in $\triangle VPT$ to show that

$$VP = \frac{\sqrt{3} \cdot r}{\sqrt{\dfrac{(5 + 2\sqrt{5})}{10}} + \sqrt{\dfrac{(5 - 2\sqrt{5})}{10}}}.$$

Hence complete the solution.

11th British Mathematical Olympiad, 1975

The 1975 paper (like the 1965–74 papers) differs in several ways from later papers. Only three hours were allowed. There were more questions — eight — than on later papers, and many of the problems are shorter and perhaps more standard than on later papers.

Moreover, the original rubric — 'Do as much as you can' — suggests that candidates were not expected to complete all the questions.

1. Given that x is a positive integer, find (with proof) all solutions of
$$[\sqrt[3]{1}] + [\sqrt[3]{2}] + \ldots + [\sqrt[3]{(x^3 - 1)}] = 400.$$

(a) Check that $[\sqrt[3]{1}] = [\sqrt[3]{2}] = \ldots [\sqrt[3]{7}] = \underline{\quad}$. Hence these first seven terms contribute a total of $\underline{\quad}$ to the LHS of the equation (∗).

(b) Find the largest N such that $[\sqrt[3]{8}] = [\sqrt[3]{9}] = \ldots [\sqrt[3]{N}] = 2$. Hence calculate how much these terms contribute to the LHS of the equation (∗).

(c) Prove that $[\sqrt[3]{y}] = k$ precisely when $k^3 \leqslant y < (k + 1)^3$. Calculate the exact contribution which the $\underline{\quad}$ terms with $k = 3$ make to the LHS of (∗). Do the same for $k = 4$. Hence find the unique value of x which satisfies (∗).

2. The first n prime numbers $2, 3, 5, \ldots, p_n$ are partitioned into two disjoint sets A and B. The primes in A are $a_1, a_2, \ldots, a_h$, and the primes in B are $b_1, b_2, \ldots, b_k$, where $h + k = n$. The two products
$$\prod_{i=1}^{h} a_i^{\alpha_i} \quad \text{and} \quad \prod_{i=1}^{k} b_i^{\beta_i}$$
are formed, where the α_i and β_i are any positive integers. If d divides the difference of these two products, prove that either $d = 1$ or $d > p_n$.

Denote the two products by Π and Π'. Suppose that $d \neq 1$ divides $\Pi - \Pi'$. Let p be the smallest prime factor of d.

(a) Explain why $p \mid (\Pi - \Pi')$.

(b) Suppose that $p \leqslant p_n$. Then $p = p_i$ (some i, $1 \leqslant i \leqslant n$) is one of the first n primes, and so occurs (to a positive power) in one of the two products; say, $p \mid \Pi$ (so that $p \in A$). Now use (a) to deduce that $p \mid \Pi'$, and hence that $p \in B$—contrary to the given fact that A and B are disjoint.

(c) Conclude that $p > p_n$. Since $d \geqslant p$, the result follows.

3. Use the pigeonhole principle to solve the following problem. Given a point O in the plane, the disc S with centre O and radius 1 is defined as the set of all points P in the plane such that $|OP| \leqslant 1$, where OP is the distance of P from O. Prove that if S contains seven points such that the distance from any one of the seven points to any other is $\geqslant 1$, then one of the seven points must be at O.

Let S be the disc $|z| \leqslant 1$ in the complex plane. Partition S into seven parts:

$$\{0\}, \text{ and } S_k = \{z \in S; z \neq 0, k(\pi/3) \leqslant \arg(z) < (k+1)(\pi/3)\} \quad (0 \leqslant k \leqslant 5).$$

Suppose none of the seven given points is equal to 0. Then one of the six sets S_i must contain (at least) two of the points. Prove that these two points will then be at a distance <1 from each other.

---

4. In a triangle ABC, three parallel lines AD, BE, and CF are drawn, meeting the sides BC, CA, and AB in D, E, and F respectively. The points P, Q, and R are collinear and divide AD, BE, and CF respectively in the same ratio $k:1$. Find the value of k.

---

(1) Let $\mathbf{a}$, $\mathbf{b}$, $\mathbf{c}$, $\mathbf{d}$, $\mathbf{e}$, $\mathbf{f}$, $\mathbf{p}$, $\mathbf{q}$, and $\mathbf{r}$ be the position vectors of A, B, C, D, E, F, P, Q, and R respectively, and let $\mathbf{x} = \mathbf{d} - \mathbf{a}$ be a vector in the direction of the three parallel lines. Choose t such that $\mathbf{d} = t\mathbf{b} + (1-t)\mathbf{c}$. We show that $k = \frac{1}{2}$.

(a) Use the fact that $AP : PD = k : 1$ to show that $\mathbf{p} = [\mathbf{a} + kt\mathbf{b} + k(1-t)\mathbf{c}]/(k+1)$. (Note that $k \neq -1$, since $k = -1$ would correspond in some sense to P being a 'point at infinity'.)

(b) The general point on the line BE has vector $\mathbf{b} + s\mathbf{x} = -s\mathbf{a} + (st+1)\mathbf{b} + s(1-t)\mathbf{c}$, and the general point on the line AC has vector $r\mathbf{a} + (1-r)\mathbf{c}$. Deduce that $\mathbf{e} = (1/t)\mathbf{a} + (1 - (1/t))\mathbf{c}$, and hence that $\mathbf{q} = [(k/t)\mathbf{a} + \mathbf{b} + k(1 - (1/t))\mathbf{c}]/(k+1)$.

(c) Do the same with CF and AB to show that $\mathbf{f} = [1/(1-t)]\mathbf{a} + [-t/(1-t)]\mathbf{b}$, and hence that $\mathbf{r} = [(k/(1-t))\mathbf{a} - (kt/(1-t))\mathbf{b} + \mathbf{c}]/(k+1)$.

(d) The vectors $\mathbf{a}$, $\mathbf{b}$, and $\mathbf{c}$ are linearly dependent. Hence, to obtain a necessary and sufficient condition for $\mathbf{p}$, $\mathbf{q}$, and $\mathbf{r}$ to be collinear (that is, for $\mathbf{p} - \mathbf{q}$ and $\mathbf{q} - \mathbf{r}$ to be linearly dependent) it is convenient to take the origin at A. Put $\mathbf{a} = \mathbf{0}$ and show that $\mathbf{p} - \mathbf{q}$ is then a multiple of $\mathbf{q} - \mathbf{r}$ if and only if $2k^2 + k - 1 = 0$. Hence conclude that $k = \frac{1}{2}$.

(e) Let ABC be equilateral and let D be the midpoint of BC. Locate P, Q, and R in this case and check that they are collinear.

(2) You might like to find a direct proof of the converse of what is proved in (1): namely, if parallel lines AD, BE, and CF are drawn through the vertices of triangle ABC, meeting the opposite sides at D, E, and F, and if P, Q, and R are the points of trisection of the segments AD, BE, and CF (with $PD = 2 \cdot AP$, and so on), then P, Q, and R are collinear.

5. Let m be a fixed positive integer. You are given that

$$\binom{2m}{0} + \binom{2m}{1}\cos\theta + \binom{2m}{2}\cos 2\theta + \ldots + \binom{2m}{2m}\cos 2m\theta$$
$$= (2\cos\tfrac{1}{2}\theta)^{2m}\cos m\theta,$$

where there are $2m+1$ terms on the LHS; the value of each side of this identity is defined to be $f_m(\theta)$. The function $g_m(\theta)$ is defined by

$$g_m(\theta) = \binom{2m}{0} + \binom{2m}{2}\cos 2\theta + \binom{2m}{4}\cos 4\theta + \ldots + \binom{2m}{2m}\cos 2m\theta.$$

Given that there is no rational k for which $\alpha = k\pi$, find the values of α for which $\lim_{m\to\infty}[g_m(\alpha)/f_m(\alpha)] = \tfrac{1}{2}$.

(1) You need to find some connection between f_m and g_m. Since $g_m(\theta)$ is equal to $f_m(\theta)$ without the 'odd' terms (in θ, 3θ, and so on), this suggests looking at

$$f_m(\theta+\pi) = 1 + \binom{2m}{1}\cos(\theta+\pi) + \binom{2m}{2}\cos 2(\theta+\pi)$$

$$+ \ldots + \cos 2m(\theta+\pi)$$

$$= 1 - \binom{2m}{1}\cos\theta + \binom{2m}{2}\cos 2\theta - \ldots + \cos 2m\theta,$$

in which all the 'odd' terms have negative coefficients. Hence

$$g_m(\theta) = \frac{[f_m(\theta) + f_m(\theta+\pi)]}{2}$$

$$= \frac{1}{2}\left[\left(2\cos\left(\frac{\theta}{2}\right)\right)^{2m}\cos m\theta + \left(2\cos\left(\frac{\theta+\pi}{2}\right)\right)^{2m}\cos m(\theta+\pi)\right].$$

(2) Show that $\cos\left(\dfrac{\theta+\pi}{2}\right) = -\sin\left(\dfrac{\theta}{2}\right)$. Conclude that

$$\frac{g_m(\alpha)}{f_m(\alpha)} = \frac{\dfrac{1}{2}\left[\left(2\cos\left(\dfrac{\alpha}{2}\right)\right)^{2m}\cos m\alpha \pm \left(2\sin\left(\dfrac{\alpha}{2}\right)\right)^{2m}\cos m\alpha\right]}{\left(2\cos\left(\dfrac{\alpha}{2}\right)\right)^{2m} \cdot \cos m\alpha}$$

$$= \frac{1}{2}\left[1 \pm \left(\tan\left(\frac{\alpha}{2}\right)\right)^{2m}\right].$$

Show that this converges (to ___) as m tends to ∞ if and only if $|\tan(\alpha/2)| <$ ___; that is, if and only if $2n\pi - (\pi/2) < \alpha < 2n\pi + (\pi/2)$ for some $n \in \mathbb{Z}$.
(3) How does this calculation depend on the given fact that 'there is no rational k for which $\alpha = k\pi$'?

---

6. Prove that, if n is a positive integer greater than 1, and $x > y > 1$, then

$$\frac{x^{n+1} - 1}{x(x^{n-1} - 1)} > \frac{y^{n+1} - 1}{y(y^{n-1} - 1)}.$$

---

(1) There is no obvious way of using standard results about inequalities here.

(a) Let $f(x) = (x^{n+1} - 1)/(x^n - x)$. Compute the numerator $g(x)$ of the derivative $f'(x) = g(x)/(x^n - x)^2$.

(b) Show that $g(x) = (x - 1)(x^{2n-1} + x^{2n-2} + \ldots + x^{n+1} - (n - 1)x^n + x^{n-1} + \ldots + x + 1)$.

(c) Use $x > 1$ to conclude that $f(x)$ is an increasing function for values of x relevant to the problem. Hence complete the solution.

(2) Alternatively, it is enough to show that $S = y(x^{n+1} - 1)(y^{n-1} - 1) - x(x^{n-1} - 1)(y^{n+1} - 1) > 0$. Show that the expression S can be factorized as $(x - y)(xy - 1)(\ldots)$. Finally, observe that the final bracket can be written as the sum of $[n/2]$ positive terms

$$(x^{n-1} - 1)(y^{n-1} - 1) + xy(x^{n-3} - 1)(y^{n-3} - 1)$$
$$+ x^2 y^2 (x^{n-5} - 1)(y^{n-5} - 1) + \ldots.$$

---

7. Prove that there is only one set of real numbers $x_1, x_2, \ldots, x_n$ such that

$$(1 - x_1)^2 + (x_1 - x_2)^2 + \ldots + (x_{n-1} - x_n)^2 + x_n^2 = \frac{1}{n+1}.$$

---

(1)(a) Let $u_0 = 1 - x_1$, $u_1 = x_1 - x_2, \ldots, u_{n-1} = x_{n-1} - x_n$, $u_n = x_n$. Show that $\sum_{i=0}^{n} u_i = 1$. Use this to rewrite the given equality in the form

$$u_0^2 + u_1^2 + \ldots + u_n^2 = \frac{1}{n+1} (u_0 + u_1 + \ldots + u_n)^2.$$

(b) Show this implies $n(u_0^2 + u_1^2 + \ldots + u_n^2) - 2\sum_{i<j} u_i u_j = 0$. Conclude that $\sum_{i<j} (u_i - u_j)^2 = 0$. Hence complete the solution.

(2) Alternatively, use the Cauchy–Schwarz inequality $((\sum_{i=0}^{n} a_i b_i)^2 \leqslant (\sum_{i=0}^{n} a_i^2)(\sum_{i=0}^{n} b_i^2)$, with equality if and only if $a_i = \lambda b_i$ for all i) with $a_i = u_i$ and $b_i =$ ____ for all i. Hence prove that $\sum_{i=0}^{n} u_i^2 \geqslant 1/(n+1)$. Conclude that there is only one set of real numbers $x_1, x_2, \ldots, x_n$ satisfying $\sum_{i=0}^{n} u_i^2 = 1/(n+1)$.

---

8. The interior of a wine glass is a right circular cone. The glass is half filled with water and is then slowly tilted so that the water reaches a point P on the rim. If the glass is further tilted (so that water spills out), what fraction of the conical interior is occupied by water when the horizontal plane of the water level bisects the generator of the cone furthest from P?

Let $\mathscr{C}$ be the cone, A its apex, and M the midpoint of the generator of the cone 'furthest from P'. Let Π be the plane through P and M representing the final water level.

(a) Shift the cone (and the plane Π) so that A is at the origin, and the axis of the cone goes up the positive z-axis. Any stretch in the z-direction changes $\mathscr{C}$ into another right circular cone, shifts M to the new midpoint of the 'generator furthest from P', and changes all volumes by the same scale factor. All possible configurations can be obtained in this way, and the answer to the problem is the same for all of them. Thus it suffices to solve the problem for your favourite cone $\mathscr{C}$.

(b) Assume that $\angle MAP$ is $\pi/2$, that $AM = 1$ (so $AP =$ ____). Then $PM =$ ____. Let X lie on PM, with AX on altitude of $\triangle APM$. Use similar triangles to show that $AX =$ ____ .

(c) Let C be the centre of the circular 'base' of the cone. Find the radius CP of the base. Hence show that the volume of the whole cone is $2\sqrt{2} \cdot \pi/3$.

(d) It remains to find the volume of the water—that is, the volume of the cone between the plane Π and the apex A. The plane Π cuts the cone $\mathscr{C}$ in an e**i**e E, with major axis PM of length $2a = \sqrt{5}$. Thus you have to find the volume of a pyramid with base E and height $AX = 2/\sqrt{5}$. If you could find the length $2b$ of the minor axis, then you could use the formula 'πab' to find the area of the base E. The volume of water would then be equal to $(\frac{1}{3}\pi ab) \cdot AX$.

(e) Show that $b = 1$. (The equation of the cone is $x^2 + y^2 = z^2$. If CP points in the x-direction, then the equation of the plane Π is $3z - x = 2\sqrt{2}$. Substituting for z gives the equation of the ellipse E as $8x^2 - 4\sqrt{2}x + 9y^2 = 8$. The minor axis runs in the y-direction, and $y^2 = \frac{4}{9}(2 + \sqrt{2}x - 2x^2)$. Since $-2 \leqslant x \leqslant 1$, $x + x^2 \geqslant \frac{1}{4}$, so $2 - x - x^2 \leqslant \frac{9}{4}$. Hence $y^2 \leqslant 1$.) Hence show that, in its final position, the volume of water is exactly $1/2\sqrt{2}$ of the volume of the glass.

Appendix
The International Mathematical Olympiad:
UK teams and results, 1967–1996

The problems in this book are intended to challenge and stimulate large numbers of interested young mathematicians in their last years at school. However, they also constitute the first round of the selection process for the team which has represented the UK at the International Mathematical Olympiad each year since 1967 (with the exception of 1980, when the IMO was cancelled, and two teams took part in alternative events). Indeed, the prospect of being invited to attend the IMO in 1967 or 1968 was perhaps the main reason for starting the British Mathematical Olympiad in 1965. It therefore seems appropriate to end with a brief record of UK participation in the IMO.

The IMO began in 1959 in Romania, and was at first restricted to Eastern European countries. Starting in the mid-1960s, countries from the West were gradually invited to take part. Teams from different countries gather each July to sit two four and a half hour papers, each with just three problems. Until 1981 countries could send a team of up to eight students. By 1982 the number of teams had increased substantially, and in that year countries were invited to send teams of up to four students. Since 1983 countries have been invited to send up to six students. The event has continued to grow: in 1967, 13 countries took part; in 1982, 30 countries took part; in 1996, 75 countries took part.

The IMO is officially an individual competition. The medals are officially known as 'first' (1), 'second' (2), and 'third' (3) prizes, but are often referred to as 'gold', 'silver', and 'bronze' medals. The proportions of each have varied slightly from year to year: at most one half of contestants receive a prize, with the ratio gold : silver : bronze being approximately $1:2:3$.

For each year we list the host country for the IMO; the UK team Leader and Deputy Leader; the UK team's position, score, and medal tally; and the team members, their medals, and their schools.

1967 YUGOSLAVIA — Team Leader: **R. C. Lyness**   Deputy: **N. Routledge**
Position 4/13;   Score 231/336;   Medals 1G, 2S, 4B

Michael J. Cullen (3)	Winchester C	Simon P. Norton (1)	Eton C
Anthony L. Davies (3)	Manchester GS	Patrick J. Phair (2)	Winchester C
David W. Garland (3)	Eton C	George Cameron Smith (3)	KEVI GS, Stratford
Robert Hill	Manchester GS	Malcolm J. Williamson (2)	Manchester GS

1968 USSR — Team Leader: **N. Routledge**   Deputy: **David Monk**
Position 4/12;   Score 263/320;   Medals 3G, 2S, 2B

Clifford C. Cocks (2)	Manchester GS	William Porterfield (1)	Westminster S
Elwyn H. Davies (3)	Manchester GS	Michael Proctor	Shrewsbury S
Noel Leaver (3)	Burnley GS	John Scholes (2)	Winchester C
Simon P. Norton (1)	Eton C	Malcolm J. Williamson (1)	Manchester GS

1969 ROMANIA — Team Leader: **F. R. Watson**   Deputy: **L. Beeson**
Position 5/14;   Score 193/320;   Medals 1G, 1S, 1B

David J. Aldous (2)	Exeter S	John F. Segal	Dulwich C
Patrick M. Bennett	RGS, Newcastle	Peter D. Smith (3)	Manchester GS
A. J. McIsaac	Charterhouse	Alan G. Trangmar	Dulwich C
Simon P. Norton (1)	Eton C	Nick S. Wedd	Kelly C, Tavistock

1970 HUNGARY — Team Leader: **B. Thwaites**   Deputy: **Margaret Hayman**
Position 6/14;   Score 180/320;   Medals 1G, 0S, 6B

Charles J. K. Batty (3)	Rugby S	David Grubb	Greenford GS
Stuart Bell	Dartford GS	John Proctor	QMGS, Walsall
Daniel Jeremy M. Edwards	RGS, Newcastle	John F. Segal	Dulwich C
Stephen B. Furber	Manchester GS	Bernard W. Silverman (1)	City of London S

1971 CZECHOSLOVAKIA — Team Leader: **Frank Budden**   Deputy: **Peter Reynolds**
Position 5/15;   Score 110/320;   Medals 0G, 1S, 4B

David J. Allwright (3)	Rugby S	David J. Jackson (2)	Perse S
Stuart Bell	Dartford GS	Nicholas S. Manton (3)	Dulwich C
Daniel Jeremy M. Edwards	RGS, Newcastle	Angus H. Rodgers (3)	Royal Belfast Acad. Inst.
Christopher R. Hills (3)	Dulwich C	Colin W. Vout (3)	Dulwich C

1972 POLAND — Team Leader: **R. C. Lyness**   Deputy: **Margaret Brown**
Position 5/14;   Score 179/320;   Medals 0G, 2S, 4B

David J. Allwright (2)	Rugby S	David J. Jackson (2)	Perse S
G. Mungo Carstairs	George Watsons C	Peter Jackson	RGS, Newcastle
David J. Goto (3)	St. Pauls S	Andrew L. James (3)	Sherborne S
Ian J. Holyer (3)	St. Benedicts S	James E. Macey (3)	Nottingham HS

1973 USSR — Team Leader: **Frank Budden**   Deputy: **R. W. Payne**
Position 5/16;   Score 164/320;   Medals 1G, 0S, 5B

Michael D. Beasley (3)	Kingston GS	David Pritchard (3)	QMGS, Walsall
David J. Goto (1)	St. Pauls S	A. J. Scholl (3)	Worth S
Guy Herzmark (3)	Highgate S	Malcolm K. Sparrow	Millfield S
John Hurley	Ernest Bailey GS	Richard J. Treitel (3)	Eton C

1974 GDR — Team Leader: **R. C. Lyness**   Deputy: **David Monk**
Position 9/18;   Score 188/320;   Medals 0G, 1S, 3B

Michael P. Allen	Woking County Boys S	Michael A. Gray	Perse S
Andrew B. Apps (2)	Kings S, Canterbury	Richard C. Mason	Manchester GS
Michael D. Beasley (3)	Kingston GS	David J. Seal	Winchester C
John E. Cremona (3)	Perse S	A. J. Wassermann (3)	RGS, Newcastle

1975 BULGARIA

Team Leader: **R. C. Lyness**   Deputy: **John Durran**
Position 5/17;   Score 239/320;   Medals 2G, 2S, 3B

Pelham M. Barton (3)	Mill Hill S	John R. Rickard (1)	City of London S
David G. Frankis (2)	RGS, Newcastle	David J. Seal (2)	Winchester C
John Peter Hartley (3)	Manchester GS	Paul Verschueren	Manchester GS
Jonathan J. Hitchcock (1)	Kingston GS	David S. Walker (3)	Manchester GS

1976 AUSTRIA

Team Leader: **R. C. Lyness**   Deputy: **Colin Goldsmith**
Position 2/19;   Score 214/320;   Medals 2G, 4S, 1B

Patrick L. H. Brooke (3)	Winchester C	Jonathan Hitchcock (2)	Kingston GS
Douglas F. Easton (2)	St. Edmunds C, Ware	Richard C. Mason (1)	Manchester GS
Simon C. Farmbrough (2)	Oundle S	John R. Rickard (1)	City of London S
David G. Frankis (2)	RGS, Newcastle	Alex J. Ryba	Manchester GS

1977 YUGOSLAVIA

Team Leader: **R. C. Lyness**   Deputy: **Terry Heard**
Position 3/21;   Score 190/320;   Medals 1G, 3S, 3B

Richard E. Borcherds (2)	KES, Birmingham	Brian A. King (3)	St. Pauls S
Andrew G. Buchanan (2)	Edinburgh Acad.	Richard Pennington (3)	Wyggeston SFC
Alan G. Bustany (2)	Sullivan Upper S	John R. Rickard (1)	City of London S
Philip E. Gibbs (2)	Currie HS	Paul M. Stickland	Dulwich C

1978 ROMANIA

Team Leader: **R. C. Lyness**   Deputy: **John Hersee**
Position 3/17;   Score 201/320;   Medals 1G, 2S, 2B

Richard E. Borcherds (1)	KES, Birmingham	Philip E. Gibbs (2)	Currie HS
Alan J. Dix (2)	Howardian HS, Cardiff	Martin P. Gilchrist	St. Albans S
Richard M. Durbin (3)	Highgate S	Richard Pennington (3)	Wyggeston SFC
Clive A. Frostick	Dulwich C	Peter P. Taylor	Dulwich C

1979 UK

Team Leader: **Colin Goldsmith**   Deputy: **Terry Heard**
Position 4/23;   Score 218/320;   Medals 0G, 4S, 4B

Nigel Boston	Manchester GS	W. D. K. Green	St. Albans S
Henry G. Bottomley	Westminster S	Colin H. A. Hogben	Kent C
Richard M. Durbin	Highgate S	Paul Taylor (3)	RGS, H. Wycombe
Clive A. Frostick	Dulwich C	Nigel C. Westbury	Canford S

1980 MONGOLIA (cancelled)

[1980A FINLAND

Team Leader: **David Cundy**   Deputy: **John Hersee**

Graham M. Clemow	KCS, Wimbledon	Colin H. A. Hogben	Kent C
Thomas Davy	Kingston GS	Jane Mills	Leighton Park S
Johnathan M. Edwards	Perse S	Malcolm Riley	Manchester GS
David J. Harvey	Bishop Veseys S	Christopher P. Smith	St. Johns C, Southsea

1980B LUXEMBOURG

Team Leader: **David Monk**   Deputy: **Graham Howlett**

Henry G. Bottomley	Westminster S	John Lister	St. Albans S
W. D. K. Green	St. Albans S	Fergus R. McInnes	Royal HS, Edinburgh
Peter B. Kronheimer	City of London S	Jeremy C. Rickard	City of London S
Bernard Leek	Alleynes S, Stone	Richard L. Taylor	Magdalen C]

1981 USA

Team Leader: **R. C. Lyness**   Deputy: **John Hersee**
Position 3/27;   Score 301/336;   Medals 3G, 4S, 1B

Mark David Bennet (1)	Ranelagh S	Imre B. Leader (2)	St. Pauls S
W. Tim Gowers (1)	Eton C	Stephen J. Montgomery-Smith (2)	Kings S, Peterborough
Ian R. Jackson (2)	Rugby S	Jeremy C. Rickard (1)	City of London S
Peter B. Kronheimer (2)	City of London S	S. Paul Smith (3)	RGS, Newcastle

1982 HUNGARY Team Leader: **R. C. Lyness**   Deputy: **David Monk**
Position 10/30;   Score 103/168;   Medals 0G, 0S, 4B

| Paul N. Balister (3) | KCS, Wimbledon | Piers J. Coxon (3) | Dulwich C |
| D. A. Chalcraft (3) | Latymer Upper S | William Sutherland (3) | Monmouth S |

1983 FRANCE   Team Leader: **R. C. Lyness**   Deputy: **David Monk**
Position 11/32;   Score 121/252;   Medals 0G, 3S, 1B

Paul N. Balister (2)	KCS, Wimbledon	Alexander S. Clark (2)	Charterhouse
Richard S. Biswas (2)	Leighton Park S	Ian J. Leary	Ormskirk GS
Mark V. Bravington	City of London S	Alison McDonald (3)	South Park SFC

1984 CZECHOSLOVAKIA   Team Leader: **David Monk**   Deputy: **Frank Budden**
Position 6/34;   Score 169/252;   Medals 1G, 3S, 1B

Paul N. Balister (1)	KCS, Wimbledon	Marcus Moore (2)	Manchester GS
Richard S. Biswas (2)	Leighton Park S	Matthew J. Richards (2)	Millfield S
Michael Harrison	Loughborough GS	Ian D. B. Stark (3)	Winchester C

1985 FINLAND   Team Leader: **R. C. Lyness**   Deputy: **David Monk**
Position 10/38;   Score 121/252;   Medals 0G, 2S, 3B

Chris Kilgour	Kings S, Gloucester	Matthew J. Richards	Millfield S
John Longley (3)	Portsmouth GS	Alex Selby (3)	City of London S
Marcus Moore (3)	Manchester GS	Ian D. B. Stark	Winchester C

1986 POLAND   Team Leader: **David Monk**   Deputy: **Paul Woodruff**
Position 11/37;   Score 141/252;   Medals 0G, 2S, 3B

James Angus	Wells Cath. S	John Longley (3)	Portsmouth GS
Kevin M. Buzzard (3)	RGS, H. Wycombe	Tom Roughan (3)	Leeds GS
Dominic Joyce (2)	QE Hospital	Andrew Smith (2)	City of London S

1987 CUBA   Team Leader: **R. C. Lyness**   Deputy: **Terry Heard**
Position 10/42;   Score 182/252;   Medals 1G, 2S, 2B

James Angus	City of London S	George Russell (2)	KES, Birmingham
Kevin M. Buzzard (1)	RGS, H. Wycombe	Andrew Smith (2)	City of London S
Gareth McCaughan (3)	Lincoln Christs Hospital S	Gerard Thompson (3)	St. Georges C, Weybridge

1988 AUSTRALIA   Team Leader: **David Monk**   Deputy: **Paul Woodruff**
Position 11/49;   Score 121/252;   Medals 0G, 3S, 2B

Colin Bell (3)	Trinity S, Croydon	Chris R. Nash (3)	KES, Birmingham
Malcolm J. Law (2)	KES, Birmingham	Oliver M. Riordan (2)	St. Pauls S
Gareth McCaughan (2)	Lincoln Christs Hospital S	Joshua R. X. Ross	City of London S

1989 FRG   Team Leader: **David Monk**   Deputy: **Paul Woodruff**
Position 20/50;   Score 122/252;   Medals 0G, 2S, 1B

Katherine Christie	Portsmouth GS	Chris R. Nash (2)	KES, Birmingham
Vin de Silva (3)	Dulwich C	Oliver M. Riordan (2)	St. Pauls S
Clive R. Jones	Dulwich C	John Simcox	Chase HS, Malvern

1990 CHINA   Team Leader: **Peter Shiu**   Deputy: **Paul Woodruff**
Position 10/54;   Score 139/252;   Medals 2G, 0S, 2B

Vin de Silva (1)	Dulwich C	Tom Leinster	Lancing C
Michael Fryers (3)	Altrincham GS	Oliver M. Riordan (1)	St. Pauls S
Alan Iwi	Westminster S	Amites Sarkar (3)	Winchester C

1991 SWEDEN

Michael Fryers (1)
Oliver T. Johnson

Robin Michaels (3)

Team Leader: **Tony Gardiner**   Deputy: **Paul Woodruff**
Position 17/56;   Score 142/252;   Medals 1G, 0S, 2B
Altrincham GS   Luke T. Pebody   Rugby S
KES,   Adam P. Shepherd   KES,
Birmingham   Birmingham
Haberdashers   Steven P. Wilcox (3)   Portsmouth GS
Askes S

1992 RUSSIA

Oliver T. Johnson (3)

Eva R. Myers (1)

Robin Michaels (2)

Team Leader: **Tony Gardiner**   Deputy: **Christopher Bradley**
Position 5/58;   Score 168/252;   Medals 2G, 2S, 2B
KES,   Karen M. Page (3)   S. Bromsgove HS
Birmingham
Streatham Hill   Luke T. Pebody (2)   Rugby S
& Clapham HS
Haberdashers   Mark J. Walters (1)   Weald S
Askes S

1993 TURKEY

Tom Fisher (2)
Alistair Flutter (3)
Catriona Maclean (2)

Team Leader: **Adam McBride**   Deputy: **Christopher Bradley**
Position 14/73;   Score 118/252;   Medals 0G, 3S, 3B
Exeter S   Alex Paseau (3)   St. Pauls S
Hills Road SFC   Luke T. Pebody (2)   Rugby S
Harrogate GS   Chuan-Tse Teo (3)   Dulwich C

1994 HONGKONG

Ed Crane (3)
Matthew Fayers (3)
Ben Green (2)

Team Leader: **Tony Gardiner**   Deputy: **Vin de Silva**
Position 7/69;   Score 206/252;   Medals 2G, 2S, 2B
Colchester RGS   Catriona Maclean (1)   Harrogate GS
Wilsons S   Joseph Myers (1)   Rutlish S
Fairfield GS   Jacob Shapiro (2)   Westminster S

1995 CANADA

Ed Crane (1)
Matthew Fayers (3)
Ben Green (2)

Team Leader: **Tony Gardiner**   Deputy: **Christopher Bradley**
Position 10/73;   Score 180/252;   Medals 2G, 1S, 3B
Colchester RGS   Peter Keevash (3)   Leeds GS
Wilsons S   Joseph Myers (1)   Rutlish S
Fairfield GS   Louisa Orton (3)   Northgate HS

1996 INDIA

David Bibby (1)

Michael Ching (1)

Toby Gee (2)

Team Leader: **Adam McBride**   Deputy: **Philip Coggins**
Position 5/75;   Score 161/252;   Medals 2G, 4S
Ysgol Rhiwabon   John Haslegrave (2)   King Henry
VIII S
Oundle S   Hugh Robinson (2)   King Henry
VIII S
John of Gaunt S   Paul Russell (2)   St. Brides HS